A Short Environmental
History Of Italy

A SHORT ENVIRONMENTAL HISTORY OF ITALY: VARIETY AND VULNERABILITY

Gabriella Corona

Translated by Federico Poole

Copyright © Gabriella Corona
First published 2017 by
The White Horse Press, The Old Vicarage, Main Street, Winwick,
Cambridgeshire, UK

Set in 12 point Adobe Caslon Pro and Lucida Sans
Printed by Lightning Source

British Library Cataloguing in Publication Data
A catalogue record for this book is available from the British Library

ISBN 978-1-874267-97-3

To my country

Contents

TABLE OF CONTENTS

Map 1

Map of Italy

CHAPTER ONE.

BEFORE AND AFTER UNIFICATION

The human environment and the natural environment deteriorate together. We cannot adequately combat environmental degradation unless we attend to the causes related to human and social deterioration. In fact, the deterioration of the environment and of society affects the most vulnerable people on the planet.

Pope Francis, *Laudato sì* encyclical

1. The transformation of the global environment

In the course of the nineteenth century, a diverse set of factors brought about a radical transformation of Europe's environment and landscape, remodelling the relationship between the human species and the environment, and deeply influencing models of social and economic organisation. One of the primary drivers was undoubtedly population growth. Improved hygienic conditions, the progress of medicine and changes in the nutritional regime caused a decrease in mortality and a corresponding increase in the population. Europe went from 120,000,000 inhabitants in 1700, and slightly over 190,000,000 in 1800, to 400,000,000 in 1900. The main upsurge was between the second half of the eighteenth century and the first half of the nineteenth. This increase largely affected northwestern Europe and, later on, eastern Europe, rather than the Mediterranean region.

As Piero Bevilacqua has explained, in preindustrial societies the main factor limiting demographic growth, aside from disease, was the low productivity of farming systems due to discontinuity of harvests, insufficient restoration of land fertility and exposure of the soil to atmospheric agents. The widespread use of a binary rotation system alternating cereals with fallow periods did not favour the restoring of nutrients and the elimination of diseases or parasites in the soil. Fertilisation was done with animal manure, which was not available year round because, due to the lack of permanent cultivation of fodder crops, livestock was taken to pastures away from the fields for part of the year. Besides, most of the fertiliser was stored in the open for long periods, which strongly reduced its nutrient content. The eighteenth century witnessed the spread of an intensive farming system using pulses in crop rotation. Pulses served as forage, favouring the stable presence of cattle on plots, and enriched the soil with nitrogen, thereby increasing yields. Between the eighteenth and nineteenth centuries, production soared in several countries – including France, Sweden, England and Denmark – eventually exceeding 2,000 kg per hectare, with peaks as high as 4,000. England, in particular, went from an average grain yield of 1,009 kg per hectare in the late eighteenth century to 1,412 kg in the early nineteenth, 1,691 kg in 1851 and 2,018 kg in 1870.

Along with these innovations in crop rotation systems, artificial methods for the restoration of fertility began to spread in the second half of the nineteenth century. Almost in the same period, formulas for chemical fertilising were developed in France, Germany and England, setting the stage for the rise of the modern fertiliser industry. Great Britain was one of the first countries to employ artificial fertilising systems. The importation of large quantities of guano from Peru – hundreds of thousands of tons between 1850 and 1870 – helped to steer English farmers towards the use of commercial fertilisers. At the end of the century, Chile soda, ammonium sulfate, sodium nitrate, potassium salts and basic slag or Thomas phosphate began to spread in England as well as other Western European countries. Farming machinery was also gradually introduced in the second half of the nineteenth century, when the agricultural labour force underwent an unstoppable decline, with a consequent increase in salaries. In 1861, six per cent of the wheat produced in Great Britain was harvested with the McCormick harvesting machine. By 1874, this percentage had increased to fifty and by 1900 to eighty. In 1882, the percentage was one in Holland, six in France, three in Germany and four in Belgium.

Changes in fertility restoration systems and the introduction of more intensive crop rotation systems were not the only factors in the modern growth of agriculture. They went hand in hand with a broad process of change in land ownership and management involving a reduction of communal land rights. Due to the rise of the enclosure movement, England was the European country where the process of dissolution of communal land rights was more thorough, starting as early as the end of seventeenth century. After England, the European countries where the process of enclosure and privatisation of common land was most successful in the first half of the century were the Scandinavian countries – Sweden and Denmark – and Prussia. In most other European countries, the decline of the commons was stronger towards the end of the nineteenth century.

The dismantling of the commons, while it did help to increase farming yields, undermined the set of rules and collective uses that for centuries had served as a safety net of sorts for the whole European continent. The collective use of natural resources had been a means to protect farmland. The rules of sharing and seasonality discouraged the depletion of resources for reasons of personal gain. This had held true in primitive societies, and continued to be the case in the modern and contemporary age in the forest regions of western Germany and Switzerland and the pastures of Austria and southern Bavaria, where common property has guaranteed efficient management, whereas individual property has proved inefficient and destructive. As regards southern Europe, one only needs to think of the imposing system of collective land use constituted by the various forms of transhumance, from short-range local transhumance to longer routes across vast areas in the Iberian peninsula and southern France, in Sardinia, in south-central Italy, and in the Balkans, where the flocks descended from the Carpathians and Transylvania towards Moldavia, Wallachia and Dobruja beyond the Danube. Collective ownership of farmland also proved efficient as a means of managing land and preserving the environmental balance, and sometimes even in terms of technical and economic results. Notable examples include the English 'common fields', the *montes* of Galicia and the Italian *partecipanze*. According to economic historian Robert Allen, the growth of agricultural productivity and part of the growth of labour productivity in England between the 1400s and the 1800s was due to an economic organisation where peasants managed their own holdings, although they remained integrated in communal land resource systems. His original thesis challenges the traditional correlation between agrarian growth and intensive technical innovation of a capitalistic nature. The only technical innovations that truly produced economic growth, Allen argues, were land reclamation and canalisation, which were prevalently carried out as collective, not individual, endeavours.

Population growth pushed people to look for new land. This caused deep changes in land organisation and environmental equilibria. From the late eighteenth century onward, the age-old practice of deforestation witnessed an unprecedented acceleration, especially in England, Germany, France and Italy. Besides meeting a growing demand in wood – which, as we shall see, went hand in hand with the growth of European industry – its purpose was to clear new land for cultivation, prevalently along the slopes of the Swiss, French and Italian Alps, and along the Apennine mountain range in Italy.

Further land was reclaimed by tilling meadows and pastures, not only in northern European countries such as Sweden and Finland, but also in the Balkan peninsula, where in the nineteenth century agriculture expanded both in the plains and in the hills, to the detriment of the pastoral economy. In eastern Hungary, after the second half of the eighteenth century, vast plains formerly used as grazing grounds were tilled and repopulated.

Land reclamation works and dam-building also won new land for agriculture. Marshland draining, the digging of canals and the building of dams allowed crops to be grown on once unproductive and often malarial land. The situation in Italy, as we shall see, was not homogeneous. In the Po River valley, major canalisation and drainage works doubled and even trebled crop yields. Along the slopes of the Apennine mountain range, instead, terracing and the building of dry-stone walls and small hydraulic works favoured the development of arboreal crops. In Holland, instead, farmland was mainly expanded by building polders to shield coastal areas from high tides.

In the short term, this conquest of new spaces offset increased demographic pressure by allowing for increased agricultural yields. In the long term, however, it often had serious repercussions for the environment. Such is the case, for example, with the plains of Hungary, where, in the second half of the nineteenth century, a vast water management operation cut off huge land tracts from the periodic floods of the Tisza, one of the main tributaries of the Danube. In the long run, this accentuated drought issues and the presence in the soil of harmful substances for agriculture.

To fully grasp the degree to which the environment and natural resources had a central role in the transformation of the West in the nineteenth century, we need to look beyond the confines of Europe, to the countries of the New World. First of all, it was in the colonies that a whole range of new agroforestal practices aimed at improving the yields of arable lands were first tested and developed. It was here that English, French and Dutch botanists and officials experimented with new crops and farming methods on different kinds of soil under different climatic conditions. Secondly, as Kenneth Pomeranz has pointed out, it was the colonies that guaranteed – thanks to products such as tea, sugar, wood and cotton – the calorie input required to support Europe's steep demographic growth, which the continent would have been unable to sustain endogenously, due to its scarcity of energy sources and natural resources such as land.

The nineteenth century witnessed an intensification and an expansion of the process which Alfred Crosby appropriately dubbed 'ecological imperialism', which had already gotten under way in the previous century with the spread of maize, rice and potato. This contributed to the modern homogenisation of the world landscape described by Paolo Malanima. European agriculture and farming landscapes changed, as did the ecosystems of vast regions outside Europe. Clive Ponting has dwelled extensively on these transformations. Actually, even before the nineteenth century, European colonisation had already transformed the land and environment of vast areas. One has only to think of the sugarcane plantations of Brazil and many Caribbean islands – where intensive cultivation caused serious soil depletion, notably in Jamaica and Barbados – or of the farming of coffee in Brazil, tobacco in Virginia and Maryland and cotton in South Carolina. However, the acceleration of these transformations in the nineteenth century was unprecedented. In the Americas, crops were introduced to new areas – cotton to Virginia,

Georgia, Alabama, Louisiana, Texas, and rubber to Amazonia – and the demands of European economic growth produced radical environmental changes. In Southeast Asia, tea, rice and rubber plantations rapidly extended to Assam, Ceylon, southern India, Malaysia, Burma, Indochina and Thailand. In Africa, the main crops were coffee and cocoa, with plantations in Malawi, Kenya, Uganda, Ghana and the Ivory Coast. The production of palm oil, used as a lubricant in industries and to make soap, flourished especially in western Africa. The growing demand for animal products determined the rise of new livestock-farming districts in the Argentinian pampas and in New Zealand. Woodcutting for exportation caused widespread forest destruction in the British Honduras, in Burma, on the Indian coast of Malabar, on the Pacific islands (Fiji, Marquesas, Hawaii) and on the Ivory Coast. There was also a European demand for products to restore soil fertility, such as guano from the west coast of South America and phosphates from Tunisia and Morocco. In some African countries, from the end of the century onward, the extraction of minerals – controlled by European companies – became a pillar of the economy. The minerals extracted included copper in Congo and Rhodesia, gold and diamonds in South Africa and iron and bauxite in Zambia and Mauritania. Their political and economic supremacy thus allowed industrialising European countries to extend their 'ecological footprint' to other continents, and thus expand the material basis of their subsistence and wealth outside Europe, causing deep and often irreversible changes to the ecosystems of these regions.

However, the most crucial factor in the development of European society and economy, and in its environmental impacts in Europe and beyond, was the sourcing of energy. Even in the Early Modern period, rising demand for energy had led to a serious shortage of wood and then to growing and increasingly widespread use of fossil coal. Employed as early as the seventeenth century for the manufacturing of glass, bricks, lead, copper and tin, in the following century coal had begun to be used regularly in iron and cast iron production. Its replacing of wood determined a change in production processes which came to be known as the 'industrial revolution', and marked an epochal transition in the relationship between human beings and the environment. As Malanima has told us, the historic passage on which modern urban-industrial societies were founded was the move from renewable energy sources – manpower, animal power, wind, water and lumber – to large-scale reliance on non-renewable ones. Agriculture, which had so far constituted the basis of all economic activities and industrial and manufacturing productions by supplying organic raw materials, gradually morphed into a sector whose exclusive purpose was to meet the demand for food.

The transition from an economy founded on renewable energy sources to a 'mineral-based' one founded on non-renewable sources was a slow process, which involved western Europe more pervasively only after the mid-nineteenth century. Nevertheless, coal production increased at an unprecedented scale, from 13,500,000 tons in 1800 to 120,000,000 in 1860 and 635,000,000 by the end of the century. This allowed a consumption of energy that would have been impossible to sustain with firewood alone. The core of the advent of urban-industrial society was thus the restriction of the function of farmland to producing food and the transition from renewable to non-renewable energy sources. A long-lasting historical phase had begun, during which raw materials were mainly procured underground and fossil energy sustained the extraordinary economic development of the West in the nineteenth and twentieth century.

2. The structural features of the Italian peninsula

Just after the unification of Italy, the country's geography, geomorphology and hydrographic regimes had not changed much for a few decades. This is not to say that they had remained unchanged through time. On the contrary, Italy's mountains and plains, hills and plateaus, rivers and lakes, and woods and coasts had undergone slow but inexorable changes due not only to geological and climatic phenomena, but also to human-induced dynamics. In his outline of the original features of the Italian peninsula, Piero Bevilacqua emphasises the early anthropisation of its environment by ancient civilisations. Etruscans, Greeks, Romans, and Arabs all left their mark on Italy's landscape and introduced new farming methods and crops. The country is strewn with archaeological vestiges of rural activities

Chapter One

Figure 1.1

The nuraghe (ancient stone building) of S. Sabina at Silanus, Sardinia. Credit: http://www.labarbagia. net/focus/storia/6915/la-civilta-preistorica-della-sardegna-prenuragica-e-nuragica-

Figure 1.2

Ancient farmhouse (*trullo*) in Puglia. Credit: http://www.italia.it/it/idee-di-viaggio/siti-unesco/ alberobello-e-i-suoi-trulli.html

that bear witness to this. These include remains of hydraulic works and ancient rural settlements, as well as the very fabric of farmland parcelling, which often has remote origins. Emilio Sereni points out that the layout of much of the Italian landscape is still influenced by the *centuriatio*, the orthogonal grid of roads and canals that the Romans used to mark out the social and agricultural space. Although every nook and cranny of Italy's natural landscape has been affected by change processes through history, these processes have accelerated to an unprecedented degree during the last two centuries, and particularly during the twentieth. This acceleration was determined by a deep change in social organisation that brought with it a different way of conceiving nature and our relationship with it.

Italy stood at the threshold of modernity not only with its rich historical heritage, but also with a particularly vulnerable landscape. High hydrogeological risk has always been a 'natural' feature of the Italian peninsula. As geologists and geographers have shown, landslides and floods have been occurring in Italy ever since remote times. The country is largely mountainous and hilly, geologically young and prevalently composed of erodible rocks and an unstable and rapidly evolving orography. Italy is bordered by the Alps to the North and crossed by the Apennines for most of the length of the peninsula. Its plains are all small – except for the Po River plain – and enclosed by hills and mountains. Mountain zones (above 700 metres) account for 35.2 per cent of the country's total surface, and hilly areas for 41.6 per cent. Plain areas, which are the most prone to marshification and flooding, encompass 23.2 per cent of the total surface. As Giuseppe Gisotti and Marcello Benedini have explained, vulnerability is accentuated by Italy's Mediterranean climate, which alternates long-lasting droughts with periods of intense rainfall. The heavy rains erode slopes, especially when they are not sufficiently covered with vegetation. The geological conformation of Italy's mountain chains makes them especially prone to landslides and floods. During the dry summer, cracks form in the prevalently clayish rocks that constitute about twenty per cent of the country's agricultural surface. In autumn and winter, water seeps in through these cracks, causing landslides and forming gullies. The soils that are at the highest hydrogeological risk are the so-called 'varicoloured scaly clays', found along the Apennine range in Liguria, Romagna and Tuscany, as well as in some areas in Basilicata, Calabria and Sicily. These clays are impermeable, so rainwater, instead of being absorbed, runs off downslope, causing impetuous and destructive floods. The siliceous sandstones constituting much of the eastern Alps and the Apennines are also prone to sliding, although not as much as clayish rocks. The Italian coastland, being composed of sand and exposed to the sea and strong sea winds, is especially susceptible to erosion.

The country's hydrographic system is also typical for a geologically young land, with watercourses having a torrential regime characterised by long dry spells and short flood phases. This regime would require torrent water to be collected in large reservoirs from which it could be drained away slowly. The new unitary state addressed the issue of water control early on, unlike other aspects of land management. Act 2248 of 20 March 1865 on public works introduced a series of regulations for hydraulic works. Through later legislative acts, the state established a Geological Bureau at the Ministry of Agriculture, Industry and Commerce, and entrusted the civil engineering department (Genio Civile) with the surveillance of riverbanks. A royal decree of 25 July 1904 set guidelines for the building of flood control works and allocated state funds to this end. Hydraulic works were addressed again in Act 774 of 13 July 1911 and Consolidated Act 1775 of 11 December 1933, which gathered legislation concerning water management and electric systems. The latter act instituted a Hydrographic Service at the Ministry of Public Works and placed the underground water of certain districts under state supervision.

So these were the environmental characteristics of Italy at the time when it went through the same major changes as all of western Europe between the eighteenth and nineteenth century, starting with demographic growth. The country's population grew from 13,200,000 in 1700 to 15,300,000 in 1750 and 17,800,000 in 1800. Today this growth of slightly above four million – fifteen more people per square kilometre – may seem modest, but for the precarious environmental equilibrium of the time, it was unsustainable. According to Malanima's calculations, if a little less than one hectare was needed for the survival of each inhabitant of the Italian peninsula, as was the case two centuries earlier, fifteen million hectares would have been needed in 1800, an amount of arable land that Italy lacked.

The strong acceleration of deforestation, which as we have seen occurred in several European countries, had an especially destructive impact on Italy's fragile landscape. No longer held back by forest cover on the mountains, rainwater ran violently downhill, flooding towns and fields. Between the late eighteenth century and unification, all the states of Italy had adopted laws and implemented policies to regulate woodcutting, keep deforestation under control and preserve hydrogeological balances. Beyond individual differences, pre-unification legislation sought to strike a balance between the safeguarding of ecological equilibria, on the one hand, and economic interests in the exploitation of wood for profit on the other. This was mainly done by favouring wood production for the market to the detriment of local community rights of use.

The first Italian forest law was approved under the government of the historic Right in 1877. It was the result of a mediation between the upholders of the protection of forest cover and those who pressed for its economic exploitation. Act 3917 of 20 June 1877 forbade woodcutting above the chestnut line, which corresponds to an altitude of about 700–800 metres. This provision was amply susceptible to interpretation, partly because chestnut trees grow at different altitudes depending on latitude, and there can hence be differences between the north and the south of the country. In fact, the law ended up favouring subjects interested in the exploitation of woodland and failed to keep deforestation in check. Only in the early decades of the twentieth century, partly thanks to advancements in silviculture, did the Italian government issue acts inspired by principles that were more protective of the woodland system. Act 3267 of 30 December 1932, entitled *Riordino e riforma della legislazione in materia di boschi e di terreni montani* (Reordering and reform of the legislation on woods and mountain areas) introduced measures to reduce hydrogeological risk in woods.

Between the last three decades of the nineteenth century and the first decade of the twentieth, the surface covered by woods and forests shrank by fifteen to thirty per cent, by about 30,000 hectares a year, giving way to crops, even on slopes and in marginal areas. This decrease was a consequence of the growing demand for wood for energy and industrial uses and, even more, of population growth in mountain areas. In 1901, it was estimated that arable land and orchards extended over 34 per cent of the Italian mountains, while only 29 per cent was covered by woodland. This shrinkage was reduced during the second and third decades of the century, when reforested areas exceeded deforested ones. Today mountain woodland has more than doubled since the beginning of the twentieth century.

In their now classic book *Le bonifiche in Italia dal '700 a oggi*, Piero Bevilacqua and Manlio Rossi-Doria have highlighted the diversity of water management issues across the Italian peninsula. Even before unification, extensive reclamation works in the plains of northern Italy had begun to transform marshy areas into highly productive farmland. These works had begun in Piemonte and Lombardy as early as the modern age. They mostly consisted of large-scale operations such as the Canale Cavour, which originates from the Po River at Chivasso, is augmented by water from the Dora Baltea river and irrigates a vast area encompassing the Lomellina and the countrysides of Vercelli and Novara. The other major irrigation work in northern Italy, the Canale Villoresi, dug in the second half of the nineteenth century, flows out of the Ticino river and crosses the Provinces of Varese, Milan and Monza, and the Brianza region.

The situation was different in the eastern zones of northern Italy. Here the part of the Po River plain grading down towards the Adriatic Sea forms lagoons in places. For centuries, the main problem in these areas had been the silting-up of rivers. The area of the Reno river, near Ferrara, is a case in point. This vast area underwent a massive land reclamation operation. Water bodies were filled in, rivers channeled and collector canals dug. The beginnings of this work went all the way back to the Middle Ages, when the local villages and farming communities were already undertaking planned water management actions. This is an exemplary case of how forms of collective land management can ensure environmental control over the centuries. Already before the unification of Italy, major land reclamation works had been started, notably a project promoted by private operators in the Valli Grandi Veronesi e Ostigliesi in the Provinces of Verona and Mantova. Thanks to the work of water management consortia and the financial support of the state, land-reclamation operations in the lower Po River plain, and particularly in the Provinces of Ferrara, Modena, Ravenna and Rovigo, were brought to completion between the late nineteenth and early twentieth century.

As to southern Italy, it was characterised by a prevalence of disseminated and scattered settlements in the hills and the mountains, since for centuries its coastal plains had been marshy, malaria-ridden, hard to cultivate and exposed to pirate attacks. Furthermore, deforestation had accelerated soil erosion and caused landslides to increase. The plains had suffered the worst effects of this process, since torrents dragged rocks and detritus into them, aggravating stagnation and swamping, especially at river mouths. Bevilacqua has shown that the land reclamation works undertaken by the Bourbon monarchy before the unification of Italy were intensive and far-ranging. Thanks to the work of the technicians and engineers of the Bureau of Bridges and Roads, Waters, Forests and Hunting, and particularly of Carlo Afan de Rivera, who directed the Bureau from 1824 onward, many areas in Campania, Puglia and Calabria were reclaimed, notably in the Vallo di Diano, along the Sarno river, in the lower Volturno river plain, in the plain of Rosarno and the Crati valley, around Brindisi and near the Salpi lake. These major projects comprised not only land reclamation and the building of a vast road network, but also an in-depth investigation of the southern landscape, which revealed that about a seventh of its surface (around a million hectares) was covered by marshes. According to Bevilacqua, through their experience in these projects, the technicians of the Bourbon administration developed a conception of state intervention which foreshadowed the integral land-reclamation plan that would be drawn up and implemented under the Fascist government. In these areas, land reclamation could not be a mere matter of draining, as in other areas in Italy. Action was required at the very source of water-related problems. The Bourbon government's plan thus also envisaged the laying out and reforesting of mountain areas, as well as actions to repopulate the plains and rebuild their infrastructures. These actions included intensification of agriculture, building roads and settlements and doing away with the *fusari* – man-modified swamps used for hemp retting, especially widespread in Campania and the Volturno plain – and replacing them with running-water retting. All this work required the financial support both of the state and of private subjects. An Act of 11 May 1855 established the 'Amministrazione generale delle bonificazioni' (general administration for land reclamation). The areas to be regenerated were divided into 46 districts, each under its own management.

Thus, some pre-unification states, such as the Kingdom of Sardinia and the Bourbon monarchy, had already shown a strong interest in land improvement and remarkable management abilities combining analysis, research and restoration actions. The unification of Italy brought a shift in a free-market direction. As regards land reclamation, the new ruling classes of unified Italy gave up the idea of land management as a means of preserving the overall ecological balance in favour of the notion that state action should strive for increases in agricultural yields and land rent. In spite of this, Act 269 of 25 June 1882 – the so-called 'Baccarini Act', entitled 'Norme per la bonificazione delle paludi e dei terreni paludosi' (Norms for the Reclamation of Marshes and Marshlands) – and subsequent acts, such as the T.U. of 22 March 1900, represented a decisive step forward in the regulation of the public sector's involvement and commitment. The state was to contribute fifty per cent, provincial and town governments 12.5 per cent, respectively, and private subjects 25 per cent, to establish hydraulic consortia and undertake 'first category' works, i.e., works aimed at simultaneously improving hygiene and agriculture. The costs of 'second category' works, instead, were to be borne almost exclusively by private subjects. The Baccarini act ended up favouring large landowners in northern Italy and did not produce significant results in the South, where an entirely different kind of state action would have been called for. Nevertheless, it was a major turning point, especially because it allowed land reclamation in the lower Po River Plain to be completed in a few decades.

The conformation of the Italian coast – which extends for almost 8,000 kilometres – had strongly influenced fishing methods. Although there were some exceptions – ports from which fishing vessels sailed out to Africa, Greece, Turkey, the Atlantic and the Canaries – there were few harbours suitable to host fleets of deep-sea fishing vessels. Usually all that was available was simple anchorages and inlets. Furthermore, as we have seen, the coasts of some southern regions were marshy and mephitic. Fishing had hence always been done prevalently inshore and venturing no further than the coasts of neighbouring regions. As in the case of forests, both the norms followed by individual communities and the legislation of the pre-unity states sought to regulate the exploitation of resources and grant their reproducibility. This was done by imposing time and technical constraints, e.g., forbidding bottom-trawl fishing during the reproductive period, or the use of poisonous substances. Down to the present day, the pursuit of individual gain has been threatening to undermine the preservation of fish stocks as a common resource.

The first post-unification norms, such as those contained in an act issued in 1877, were still conservationist, in spite of pressure for more freedom from the stronger fisher groups. It was not until 1904, 1909 and the Fascist period – with the Micheli Act of 1921 and a consolidated act of 1931 – that more permissive norms were issued and legislative tools aimed at modernising fishing were deployed. Henceforth, mechanised fishing began to take hold and the sector expanded, thanks to the use of larger boats.

Geological fragility, the vulnerability of the hydrographic system and the Mediterranean climate are not the only structural factors in Italian environmental history. Volcanism and seismicity also played a role. Italy stands out in the European continent for its many volcanoes, including still fully active ones such as Etna and Stromboli, and dormant or only moderately active ones such as Vulcano, Vesuvius, the Phlegraean Fields and the many underwater volcanos in the Tyrrhenian Sea and the Channel of Sicily. Furthermore, over the centuries the whole Italian territory, except for Sardinia, has known highly destructive earthquakes. Today, over 48 per cent of Italian population lives in a seismic zone and 35 per cent of Italian

Chapter One

Figure 1.3

Mount Vesuvius: a volcanic eruption at the foot of the mountain, 1760-1761, causing the destruction of the land and property. Coloured etching by Pietro Fabris, 1776, after his drawing, 1760-1761. Wellcome Trust.

Figure 1.4

Poggioreale. Ruins after the 1968 Belice earthquake, Sicily. Photo: Stefania Bonura.

towns are classified as being at seismic risk. The last five centuries have witnessed 174 seismic disasters, with an average of one destructive earthquake every three to four years. The most affected areas are in southern Italy, particularly Calabria and Sicily, with an average frequency of seventeen to nineteen years. The towns in southern Italy that have been totally destroyed by an earthquake in their history number 400.

The issue of how to deal with seismic events and reconstruction in their wake has always been a complex one, which has elicited a multitude of responses reflecting local differences in context, administrative culture and availability of economic resources. For many centuries, the cost of reconstruction was borne by individuals. Public administrations sometimes gave a hand by offering tax deductions or granting loans. There were also some actual state reconstruction plans, usually limited to urban planning and the control of prices and land sale. Some of the best known of these plans were implemented by the Spanish government in the Val di Noto in eastern Sicily after the earthquake of 1693, and by the Bourbon monarchy in Calabria after the earthquakes of 1783. Further such plans were enacted by the government of the Papal States after the earthquakes of Marche and Romagna in 1781 and the earthquake of the Valnerina in Umbria in 1859, and by the Kingdom of Naples after the earthquake of 1857 in Basilicata. The unification of Italy marked a halt or possibly even a rollback of structured public policies and of economic and administrative actions, at least until the last decade of the nineteenth century, when public institutions and the seismologist community began to work more closely. But the state only started to bear part of the costs and plan broader and more comprehensive reconstruction actions in the wake of the 1908 earthquake of the Straits of Messina and that of Marsica of 1915, as well as those that hit vast areas in the Appennino tosco-romagnolo in 1917 and 1919. After World War II, all the costs were assumed by the public sector, notably in the case of the earthquakes of Belice in 1968, Friuli in 1976, Irpinia in 1980, Umbria in 1997, l'Aquila in 2009 and Emilia in 2012. In some cases, public action was managed in controversial and unprincipled ways, involving corruption and a distorted use of state money. This was especially true in the case of the earthquake of Irpinia, for which a parliamentary investigation committee was established in 1989.

The history of earthquakes is one of death and destruction, incertitude and precariousness, extremely high economic and social costs, shattered communities and perturbed cultural identities, uprooting and the severing of bonds. It is also a history of migrations, moves and reconstruction. Italy has proved largely incapable of mitigating the destructive impact and the consequences of earthquakes. This should have been done by taking adequate account of the country's geological vulnerability and high seismic risk in the designing of housing, roads, production facilities, etc. Overall, earthquakes and the state's inability to implement preventive policies constitute a foundational trait of the history of Italy and of its land management.

3. The response of farming systems

Italian local populations reacted in different ways to the changes in the use of natural resources from the mid-eighteenth century onward.

In northern Italy, the dominant farming system revolved around a particular kind of farmhouse called the *cascina*. The system resulted in field arrangements that are sometimes still recognisable today. It spread across the irrigated plain spanning Lombardy and Piemonte between the Dora Baltea, Oglio and Po rivers, extending south of the Po in some areas (for example the Oltrepò Pavese and the plains of Piacenza and Parma) and eastward all the way beyond the Mincio river. The *cascina* was characterised by its closed courtyard structure. It was an imposing square farmstead whose longer sides could be over a hundred metres long, In the middle was the threshing floor, where the harvest was processed and whence it was distributed. All around were houses, large stables and barns, and storerooms and sheds. *Cascine* were first established in the 1500s. During the eighteenth century, they morphed into capitalist businesses, with an entrepreneur – the tenant or owner – permanently hired workers and relatively intensive investments.

Figure 1.5

Rice weeder northern Italy from the film *Bitter Rice* (1949). Public domain photo from Wikipedia.

The pace of the spread of this agrarian system was different from one area to another. It arose early on in the area between Adda and Ticino. Further to the east, in the Novara and Vercelli areas, it reached full bloom in the nineteenth century. This differentiation went hand in hand with the expansion of two main farming strategies: rice growing, especially in western northern Italy, and wheat and livestock farming. This evolution was made possible by the renovation of the hydraulic regime and the introduction of forage pulses, which allowed the rise of a stabled livestock industry. In the so-called *area del latte* ('milk area') in the Provinces of Cremona, Brescia and Lodi, farms increasingly required a permanent work force. The traditional agropastoral equilibria based on transhumance, which used to bring shepherds with their flocks down from the surrounding mountains, were undermined. Rice-growing fed a very intense demand for labour, but one that was concentrated in certain periods of the year. Thus, rice-paddy areas saw a prevalence of temporary labourers and woman rice-weeders (*mondine*) who worked only seasonally, migrating in from other areas.

In the *cascina* system, field working went hand in hand with rural industrial activities such as silk growing and spinning, and the working of hemp and linen. The *cascina* region was blessed with a fertile soil and an abundance of water, thanks to high rainfall and the presence of large rivers. Furthermore, its ecosystems were less vulnerable than those of the Mediterranean areas of Italy. The region's main environmental problem was water regulation, which was achieved, as we have seen, through large-scale hydraulic works. The *cascina* system was characterised by strong investments in soil productivity and an intensive use of human and animal labour. Its growth allowed for increased production capability without undermining the reproducibility of resources. Easy applicability of technical innovations also played a role in this success. The region's level ground and abundance of water facilitated the spread of mechanised farming and chemical fertilisers. In the plains irrigated by the Naviglio Grande and the Naviglio Pavese, the southern part of the Bergamo plain, the farms of the Cremona area, the western Lomellina and the rice plains of the Mantovano between the Mincio and Po rivers, machines were employed as early as the beginning of the twentieth century for all the principal farming operations. Technical innovation favoured a higher increase in crop yields than in the rest of the Italian peninsula. At the same time, heavy mechanisation and intensive use of chemical fertilisers caused serious environmental damage, which became more evident after World War II.

In northeastern Italy, as well as large areas in central Italy, a very different system from the one centred on the *cascina* had been taking hold ever since the early modern age, called the *fattoria*. This system became especially widespread from the mid 1700s onwards, particularly in Veneto, Emilia, Tuscany, Marche and parts of Umbria and Friuli Venezia Giulia. *Fattorie* were middling to large and sometimes very large estates constituted by a number of holdings with farmhouses, each entrusted to a peasant family. The workforce did not consist of day labourers, but of sharecroppers or, more rarely, tenants. In the sharecropping system, the owner supplied the land, capital, livestock, seeds and equipment, and the sharecropper his labour and all or part of the tools, as well as advancing half of certain seeds. The crop was usually shared equally. Holdings were productive units that provided sharecropping families with a certain degree of self-sufficiency. While monoculture prevailed in the flatlands of the *cascina* zone, in central Italy the sharecropping system and the presence of hills favoured mixed farming, alternating tree crops (olive, grapevine), vegetable gardens and grain fields.

Hilly areas were suited to mixed cultivation because of their sloping conformation and granite or limestone foundations, which enrich the soil with minerals. In these areas, the main environmental issue was water management, and particularly the regulation of the course of rivers. Human beings therefore terraced slopes and built rational water drainage systems. This work was done over areas of limited extension, mainly through the daily toil of peasant families who provided for inspection and maintenance. The reaction of this agrarian system to the new change dynamics was based on making the most of human energy, the intensification of family work and investments in sharecroppers' homes.

There also was some technical innovation, although it was slower than in the *cascina* area, notably the replacement of the old wooden ploughs with iron ones and the use of perphosphate. Mechanisation was slowed down by the hilly nature of the region and the small size of plots. In areas where the *fattoria* system prevailed, the negative effects of the new intensive farming methods were also not long in making themselves felt. For example, mechanised cultivation's replacement of the traditional 'slope-contour' crop system, which helped to regulate runoff, had serious impacts. After World

Figure 1.6

Masseria Torcito di Cannole, southern Italy. Public domain photo from Wikipedia.

War II, the economic crisis of the Italian mountains and hills and the migration of their population towards the plains deprived those areas of the constant presence of peasant families, which for centuries had guaranteed land maintenance. The consequence was an increase in erosion and landslides.

In the plains of south-central Italy, instead, a farming system prevailed that has been called 'agriculture without houses'. It flourished in large plains such as the Maremma, the Roman countryside and the Tavoliere, as well as smaller ones like the Maremma Pisana, the littoral of Marche, the Teramo and Vasto areas, large areas in Terra di Lavoro, the Sele river valley, the Ionic littoral and the marquisate of Crotone. This system was characterised by large estates, extensive cultivation, a migratory workforce and scarce investments. The absence of houses was due to the fact that the workforce did not reside at the farm, as they did in the *cascina* and *fattoria* systems. There were indeed farm-steads – the *tenuta* in Tuscany, the *casale* in the countryside of Rome and the *masseria di campo* in Puglia. These comprised a few houses and stables, and acted as administrative centres of sorts, where the transfer of labourers and work operations were organised and coordinated. In these areas, towns mainly served as residences for farm labourers, to such an extent that they earned the name of 'agrotowns'. The costs of this form of settlement were very high. One has only to think that the workers took days to journey from their place of residence to the fields, with obvious repercussions for productivity. This farming system was founded on an alternation between extensive cereal agriculture and transhumant grazing, and thus involved an intimate link between the mountains and the plain. The flocks were led down from the moun-tains along ancient sheep-tracks to their winter pastures and then back

Figure 1.7

Village clinging to a hill. Calabria. Source: http://www.identitainsorgenti.com/touring-club-i-borghi-da-scoprire-le-new-entry-tutte-al-sud-tra-puglia-e-calabria/

up to their summer ones. The system thus involved two forms of migration, of livestock and of farm-workers.

Houseless agriculture was partly the result of a process of adaptation of the populations of vast areas in south-central Italy to environmental issues and the precarious ecosystemic balances of this region, whose settlement was made difficult by the extent of mountainous land and the scarcity of plains, the torrential character of watercourses, the landslide-proneness of slopes and the prevalence of swampy and malarial plains. These problems were further accentuated by deforestation and tilling.

This is what Rossi-Doria called 'barebone land' (*terra dell'osso*), latifundium areas where very little capital was invested, and when it was it was almost exclusively destined to pay the workforce hired periodically to carry out the main farming chores. Here we find neither the high capital investments of the *cascina* system nor the high labour investments of the sharecropping system, far less the intensive employment of capital and labour of northeastern Italy. Both funding and building investments were scarce. The result was an elementary organisation, capable of rapidly changing crops to adapt to changed circumstances or needs. The introduction of mechanisation was due to the shrinking of the workforce and rise of salaries as a result of the late nineteenth and early twentieth century transoceanic emigration. Mechanisation spread easily into the coastal zones of Abruzzo and Molise and all the way to the Tavoliere, the Materano, the plain of Sibari, the Crotone area and the Sicilian latifundia. In the decades that followed, the employment of chemical fertilisers was also intense, although in the Italian South these may not have produced the wished-for effects, especially in the most drought-stricken years.

The rural history of the Italian South, however, does not coincide with that of the latifundium. As Rossi-Doria puts it, there was also some 'meat' on the 'bone'. Small peasant holdings were well integrated into the system. The holdings were often granted by latifundium owners themselves, under various regimes: rent, small property or sharecropping. Holdings only granted a partial income to peasants, who complemented this by working as labourers for the latifundia. This system was in place in areas where intensive farming prevailed:

olive growing in the hills of Abruzzo and the Calabrian Provinces, in the areas of Salerno, Otranto, and Bari, and in the Provinces of Palermo and Trapani; wine growing in Puglia, especially in the Province of Bari, and in Sicily, at Vittoria, Marsala, Trapani, and Catania; citrus growing in Sicily and Calabria; and garden-vegetable growing in Campania, especially around Naples. In some areas of the Italian South, arboriculture became a high-yield and high-income activity, independent of the latifundium. Not all intensive crops employed the same production model. While olive trees are suitable for dry farming, citrus trees need special kinds of plots called *giardini* (gardens); they are found in the Sorrento peninsula, along the northern and southern coasts of Sicily – especially around Messina, Palermo and later Catania – and in the part of Calabria extending up to the northern limit of the Province of Reggio. Piero Bevilacqua has shown that in this form of cultivation water, far from being an agent of de-

Figure 1.8

Orange grove, Sicily. Source: public domain image from Wordpress free gallery at http://www. fableswedding.com/chiedilo-tra-i-profumi-della-sicilia/

terioration and destruction, has been the lever of a transformation making the most of environmental resources through a sophisticated system of canals, wells and filtering tunnels. In this area – where silk growing was widespread until the eighteenth century – a tradition of advanced agriculture arose. This kind of farming differed from grain-growing because it required high capital and labour investments and a radical transformation of land and landscape.

Only after World War II, but with significant precedents in the 1930s, did the complex situation of the southern Italian countryside finally come to a turning point, thanks to a huge and extraordinary state-financed operation.

Figure 1.9

Sorrento lemons. Source: Wikipedia, Brad Coy, CC BY 2.0.

CHAPTER TWO.
THE TRANSITION TO MODERNITY

1. The decline of the commons

Over the last few years, commons have featured prominently in the debate on the environmental crisis as a means to counter the destructive tendencies of economic individualism and preserve the collective and social use of resources. Commons, which still exist in different forms in various parts of the world, are inspired by a different rationale from that of capitalist societies founded on full and exclusive property rights. The core idea behind commons is that shared use can grant more benefits than individual use by protecting natural resources and ensuring more equal distribution of revenues. Today, commons are valued for their ability to combine ecology and economy, use of natural resources and protection of environmental equilibria, social equity and collective wellbeing. This is something that modern market economies, and in many cases even public policies, have not been capable of achieving.

In the case of Italy, when one speaks of commons in an historical perspective what is meant is the collective property of communities and associations that were granted the power to manage it. Common land rights also implied a non-exclusive conception of access to resources and an open conception of possession.

In Italy there existed a diverse and elaborate array of forms of collective property and common land rights. The *regole*, widespread on the Dolomites and especially in the Cadore, were exclusively family-based. Common property was managed by groups of coheirs descending from a single progenitor. The *regola* was the assembly of coheirs. In the *società degli originari* (societies of old-stock families) – especially widespread in Lombardy and Veneto – only members of the oldest and most powerful families were admitted to the redistribution of the revenues of common property. The *vicinie* of the east-central Alps, the *comunaglie* of Liguria and the *comunanze* of the Umbria-Marche tracts of the Apennines – about 360 existed at the beginning of the twentieth century – had a community structure based on assemblies attended by all or some of the inhabitants of a town. The managing of common property and the sharing out of its revenues was the prerogative of the heads of families or households holding the right to participate in the assembly. (In the case of *vicinie*, this assembly was known as the *assemblea dei terrazzani*, that is, the 'assembly of the terrace holders'.) In other cases, the community consisted of a network of neighbours or parish members. In *partecipanze*, which were mainly concentrated in the Bologna area, access to land shares was usually granted to all residents. *Beni ademprivi*, the commons of villages in Sardinia, also fall within this category. As late as 1874, villages were assigned 186,294 hectares of land as *beni ademprivi*.

Access to commons could also be conditional on the commoners' occupation. This was the case for the *società della malga* in the central Alps, which gathered livestock owners; the *università agrarie* in Lazio, made up of owners of at least one pair of plough oxen; and the *generalità dei locati* or *università de'padroni di animali*, constituted by livestock owners who migrated along sheep tracks to winter their flocks in the Tavoliere. Membership of these associations granted a series of collective rights on the land along the migration route (grazing, wood gathering, access to water, etc.).

In southern Italy, the prevalent form was universal or communal commons – one of four types existing there, also including royal, feudal, and ecclesiastic commons. They revolved around the *universitas*, a form of organisation that became widespread with the introduction of the French administrative system. Universal commons were inalienable, since it was assumed that they belonged to the people of the *universitas* since time immemorial. Each town administration managed them in various ways: by granting access to them free of charge or subject to payment of a tax called *fida*, by giving out emphyteutic leases for a certain number of years or by renting them out. Universal commons were prevalently found along the slopes of the Apennines in Abruzzo, Sannio, Campania and Lucania, mainly above 500 metres of altitude. Out of a total of 658,000 hectares of town commons, 418,000 were in the mountains, 163,000 in the hills and 77,000 in the plains.

The transition to modernity

In the Italian South, the history of the different forms of commons between the late eighteenth and early twentieth centuries is interwoven with that of the crisis and disintegration of the feudal system, which had given rise to the latifundium system and was based on an extremely flexible relationship between allodium and commons, that is, between owned land and inalienable land placed under a number of rules, including common access rights. These rights could apply to the whole extension of the fief, according to the old precept *ubi feoda, ibi demania*. Open and closed land, property and possession, common areas and estates, and collective and exclusive rights often coexisted within the same space, in a tangle of uses and customs.

Common rights on land other than collectively owned estates were not exclusive to southern Italy. In the eighteenth and nineteenth centuries, they still applied in vast areas elsewhere in the peninsula, as in the case of summer grazing rights in flatlands in several Provinces in Veneto, in the Province of Turin and in Lazio and Tuscany (Massa, Pisa, Livorno, Grosseto). In other flatlands where capitalist intensive agrarian systems, such as the *cascina*, were taking hold, communal land rights had already disappeared in the course of the early modern age. *Pensionatico*, which was widespread in Veneto, consisted of the right to graze sheep on somebody else's land. *Vagantivo* was also very common in Veneto, meaning the right to wander freely across valleys and marshes to hunt, fish and gather reeds and other marsh plants. According to early twentieth century data, common rights in land existed in 235 towns in the Provinces of Brescia, Bergamo, Como and Sondrio, seven in the Province of Verona, 220 in the Provinces of Novara and Turin and fifty in Liguria, mainly concentrated in the Provinces of Savona and Genoa. In the former Papal States (Emilia, Lazio, Marche and Umbria), before the law of 1888 an area of 595,293.3 hectares was under common rights.

Between the modern and contemporary periods, these forms of collective property were strongly affected by the evolution of the market economy. Not to mention cases of actual usurpation or illegal property claims, whereby town lands ended up being incorporated into large estates or high-yield farming businesses. In the countryside of Bologna, vast tracts of *partecipanze* land were permanently incorporated into shareholdings, contributing to the growth of large-scale capitalist land rent. In the course of the nineteenth century, common rights were further undermined by the expansion of industrial activities such as wood and reed cutting, copper extraction and the manufacture of vitriol and magnesium sulphate. Privatisation of land, however, did not always imply the disappearance of collective property. In many cases, associations of commoners lived on even as land was being closed off, and different forms of possession coexisted peacefully.

Italian commons were certainly anything but idyllic realities. On the contrary, commoners' associations were wracked by harsh internal strife, often featuring opposing factions. Commons were the basis not only of individual wealth, but also of the social and political careers of their administrators. In spite of this, they granted protection and stability. The various forms of collective property in Italy enforced norms for conserving resources, guaranteeing their reproducibility and preventing their depletion. For example, livestock not belonging to the community was not allowed into the community's territory in the winter, when grazing was scarce, but only from the beginning of spring. In the case of freshwater fishing rights, community norms were especially concerned with the interdiction of methods that could damage fish stocks, such as tub trawls. As regards wood, no family could gather more than the quota set by the assembly. The right to gather wood had to be proportionate to the productivity of woodland. Access to woods was usually subject to a tax proportional to the quantity and quality of the gathered wood. In the case of *regole*, certain woods, called *gazi*, were set aside for the needs of the community. They were used as a reserve of wood in case of fire or served as a hydrogeological defence. As to construction lumber, only as much as necessary could be harvested, subject to formal request to the administrators of the *regola*. In some areas, trees could be cut for firewood only in coppice woods. Cutting in high stands was forbidden; on the contrary, there was an obligation to plant new trees there. It was also forbidden to graze livestock in seeded fields, in natural meadows before the harvest and in coppice woods after cutting. Another restriction was the obligation to ring pigs' noses to keep them from rooting up the ground in pastures.

A number of strict rules applied to land access by 'foreigners' and neighbours. In the *società della malga*, community statutes rigidly regulated access to pastures according to the season, to whether the livestock was large or small and to whether the shepherd or herdsman was a community member or not. Some historians believe that the need for environmental protection was at the very origin of the rise of commons. *Comunanze*, for example, were allegedly formed to prevent depletion of forest resources due to demographic growth and the expansion of cultivation. The *partecipanze* of Emilia arose as a means to keep up collective land management through a time of population growth and new settlement. At the beginning of the modern age, vast areas in the Bologna plain needed land reclamation work requiring capital investments. At the same time, local communities did not want this work to be carried out by private subjects who might thereby claim possession of common land. This led to a transition from simple common land rights to a system involving the assignment of a given extent of land for a certain number of years under *ad meliorandum* emphyteutic contracts.

The rationale behind norms forbidding the division, sale or privatisation of common land lay in the advantages the commoners derived from shared and coordinated use of resources compared to private and individual use. Privately owned land would have been inadequate for a number of necessities, such as the procuring of firewood for domestic use or manufacture, summer or winter pasture for work animals or free grazing livestock, fallow agriculture or seeding, tree-growing, or simply gathering fruit, fishing, watering animals, providing water for irrigation or domestic uses, etc. In Sardinia, the zoning of common land to meet these different needs put a very peculiar stamp on the landscape, which is still perceptible today. Sardinian commons (*beni ademprivi*) were arranged in concentric circles around a settlement. The first zone was made up of fenced or otherwise delimited plots used for intensive cultivation. The next zone was for meadows called *pardu* or *minda*, where work animals were grazed. A third zone, called *aidazzone*, was for cereal growing. After the agrarian year was over, the cultivated land was used for grazing (*pe-berili*). Together, the *pardu* and *aidazzone* constituted the *ademprivi*. They were owned by the village and their use was carefully regulated. Grazing rights applied in woods, bushy pastures and unproductive land. Seeding was forbidden here, except as a special concession.

In southern Italy, the privatisation of common land, which went hand in hand with deforestation and the expansion of grain crops, increased hydrogeological instability along the slopes of the Apennines. In Sicily, the great issue of drought and water scarcity is historically connected to the parcelling out of common land under a quota system. The privatisation of woods altered natural equilibria, transforming water-rich rivers into torrents with irregular courses and inconstant flow rates. This led to the drying up of springs, alteration of the rain regime and absorption of water by the now sterile ground. Even back when the Acts of 1806 and 1808 for the abolition of feudality were being enforced during the French occupation of Italy, it became clear that the country's farming economy would not benefit from the abolition of all forms of collective land ownership. Thus, administrations allowed the parcelling out of town common land belonging within the fields and meadows – i.e., cultivable land – category, but forbade that of woods, pastures and flooded and mountain land, which was not cultivable and not suited for individual ownership.

In redistributing commons, the use of resources was carefully regulated. In some cases, the produce was divided annually among the families. In others, milk was divided among community members proportionally to the number of animals they owned. Arable land was parcelled out according to a uniform and compulsory rotation system, for brief periods – for example, five or nine years – or sometimes even very long ones – for example, 99 years. The land was shared out according to the needs of the families. If one of them ended up possessing more land than it actually needed, the assembly of representatives could oblige it to relinquish the part in excess. Work methods and work schedules were also placed under a number of rules.

For some decades, as noted above, commons have constituted an important aspect of the environmental question and over time

The transition to modernity

Figure 2.1

Tuscan landscape. Credit: public domain photo from https://sidebright.wordpress.com/2015/03/30/
paesaggio-toscano-una-good-news-da-52-pagina/

they have acquired a positive significance. For a long time, however, in the context of the more general 'land ownership question' that accompanied the construction of nation states in the Western world, they were regarded as a hindrance to be done away with. Economists and reformers believed that capitalist exploitation of land, water, woods and the subsoil would allow for an increase in production and improve people's living conditions. The suppression of communal land rights, however, was not motivated only by economic considerations; social responsibility was also a concern for the advocates of absolute right to land ownership, which was one of the hubs of the political project of the moderate middle class that was the principal actor in the Italian Risorgimento. This emerged clearly in the course of debates held in preparation for the drawing up of the civil code of law. Commons were depicted as something that not only undermined the juridical and economic order by keeping vast areas outside the market, but also subverted the moral order and the public peace due to the conflicts arising from uncertain ownership.

The rise of a modern ownership regime and the dismantling of all communal bonds were the final results of a process that had begun in the years preceding the unification of Italy. As early as the end of the eighteenth century and the immediately following decades, a series of acts were issued to allow landowners to abolish grazing rights on their lands by paying an indemnity to *comunità* in Veneto, Tuscany, the Papal States and the Kingdom of Naples. After unification, two acts issued in 1866 and 1867 put on sale vast extents of ecclesiastic land, commons and town land. In southern Italy, the parcelling out of commons and their conversion to properties freed from collective obligations had begun with the acts for the abolition of feudality in the early decades of the nineteenth century, and had continued after unification.

Fascist legislation carried on in the same direction. Under Act 751 of 1924 on the reorganisation of commons in the Kingdom, later incorporated in an act of 1927, all existing common land rights were abolished and an effort was made to free private land still placed under age-old obligations. The landowner would cede to the town or association an amount of land that varied according to the nature of the common rights and the value of the land placed under them. As to the land already owned by communities and towns, it was to be parcelled out among the families of the original users, privileging the less wealthy. Woodland belonging to communities or associations of farmers was made public and placed under the control of forest authorities. Its use was restricted and it was declared inalienable. The use and transformation of woodland was regulated by a forest act issued in 1923, approved by Royal Decree 3267 and *Regolamento* (code) 1126 of 1926, which retained certain mechanisms for the protection of commons. Concessions to commons are also found in the act on mountains of 1952, which ruled that family *comunioni* in mountain areas could continue to manage their land according to their statutes and customs.

The elimination of the various Italian forms of common property was not a linear process. The trend to privatisation that had dominated Italian economics and juridical debate was actually reversed in the late nineteenth century as a reaction to a long and laborious process that spawned Act 5849 of 4 June 1888 on the abolition of commons in central Italy, in enactment of a ministerial project aiming at the abolition of rights to graze livestock and gather and sell grass without compensation for the former holders of these rights. On this occasion, there was a harsh confrontation between the abolitionists, who represented the interests of landowners in Parliament, and a group of jurists who, ascribing to common land rights the character of natural rights, claimed that the former users should be granted the possibility to appropriate part of the redeemed land as compensation. These disputes stimulated the flourishing of historical-juridical studies on the origin and nature of town property and commons in Italy. Alongside the privatising trend, a new one arose which regarded collective ownership as necessary in cases – such as the management of woods and water – where privatisation had caused very serious damage to the land and the economy as a whole.

Those were also years when in-depth studies were conducted on land and its management. They were years when major ministerial investigations were carried out, which revealed the variety and vitality of forms of collective appropriation of resources in Italy. The treatises and data collections commissioned by the State in the late nineteenth and early twentieth century highlighted previously unnoticed aspects of commons and their social function. The acts of the 1880s Jacini investigation on Italian agriculture, for example, include analyses of historically rooted local communal forms of land ownership and management. In his report on the Marche Region, for example, Ghino Valenti – an economist and high public official – explained how the very nature of the mountain environment and the silvopastoral economy called for regimes of land ownership capable of guaranteeing rotation in woodcutting and grazing. It was a matter of giving the word 'property' a plural meaning and adapting it to the objective needs of production and distribution. Collective property, argued Valenti, did not negate progress but allowed for associative and cooperative forms of land management, being in itself a cooperation of sorts.

Figure 2.2

Hut on Monte Motta, Sestriere, Piemonte. Public domain photo from Wikipedia.

Legislation concerning common land rights in Italy issued between the late eighteenth century and the 1950s can be seen – with some important exceptions – as an attempt to incorporate commons into the private/public, market/State dichotomy. This approach was eventually challenged on environmentalist grounds. The negative consequences of the suppression of a common conception of rural property were particularly evident in mountain and wooded areas, whose situation worsened in the second half of the twentieth century due to the crisis of mountain agriculture. Thus, commons came to be viewed increasingly as a form of economic-social management capable of harmonising economy and ecology, production and maintenance.

In spite of the great efforts made during Fascism and resumed after World War II to do away with commons, many of the abolitions envisaged by the act of 1927 were never actually carried out. In the second postwar period, an estimated tenth of the Italian peninsula, largely comprising areas in the Alps and the South, was still under common rights. In Lazio, by 1977 the common land of only 39 out of 378 towns had been privatised. In recent times, collective property was still quite extensive in both Marche and Umbria. In Lazio, again, as late as 1994 thousands of citizens were still in an irregular position due to failure to implement the Act of 1927. A strenuous resistance has thus been put up by an ancient but still vital institution founded on a world view based on solidarity. This institution can apparently still play a role even in a modernised and technological society like that in which we live today.

2. The city as ecosystem

One of the most visible effects of the major transformation undergone by European society between the eighteenth and the nineteenth century was undoubtedly the growth of cities into complex ecosystems characterised by great voracity towards resources (water, food and fuel) and an equally great waste production (sewage, garbage and polluting gases). Population growth and the development of industries led to a worsening of sanitary conditions in cities. In spite of the presence of public fountains and in some cases also of early aqueducts from which water could be drawn for private consumption, the per capita water availability was extremely scarce. Laundry was washed at public washbasins or along the banks of rivers or torrents. Only a privileged few had running water at home. Excrement was gathered in cesspools that had to be periodically emptied out. Cities were dirty and unbearably smelly, and their hygienic conditions were precarious. In London, the largest city in Europe, which doubled its inhabitants in the first half of the nineteenth century, epidemics broke out regularly, claiming thousands of lives. In 1848, the Public Health Act was issued in England. It provided for the creation of sewage infrastructure and the extension of the drinking water supply system. Town governments were granted ample powers in hygiene protection and building regulation.

The hygienist culture that took hold in the second half of the nineteenth century brought with it increasing concern for the sanitary consequences of urban pollution caused by domestic and industrial waste, sewage water and gases. This concern led to the gradual disappearance of typical diseases of premodern societies, such as typhus, cholera and dysentery.

In Italy, overall environmental conditions in towns changed similarly to those in other European countries. The Italian population grew by 12,000,000 units in five decades, from 29,000,000 in 1881 to 41,000,000 in 1931 (and later 47,000,000 in 1951). This increase mainly occurred in urban areas. Between 1861 and 1936, the population of cities in north-central Italy grew from 1,736,928 to 7,857,967. Turin, for example, went from 173,305 to 629,115 inhabitants, Milan from 267,618 to 1,115,768, Genoa from 242,447 to 634,646. In the South, Naples – which at the time of the unification of Italy had been the most populous town in the country, with 484,026 inhabitants – by 1936 had 865,913, and Palermo went from 223,689 to 411,879. The resident population of towns with a population of more than 20,000 increased constantly, by 23.7 per cent in 1881, 28.1 per cent in 1901, 35.5 per cent in 1931 and 41.1 per cent in 1951.

Although belatedly compared to European countries such as England, France and Germany, in Italy, too, the most progressive among local administrations started to take measures to cope with the effects of urban growth, especially in the north-central regions of the country and in some towns of the South, particularly Naples. A hygienist culture spread thanks to figures such as Luigi Pagliani, Alfonso Corradi and Arnaldo Cantoni. In 1879, the Società italiana d'igiene was founded in Milan. It devoted itself to carrying out studies and formulating proposals to improve the precarious sanitary situation of the poorer part of the Italian population, which was particularly exposed to malnutrition, malaria and pellagra in the countryside; and to anaemia, consumption, rickets and typhoid fevers in towns.

During the four decades spanning the turn of nineteenth and twentieth centuries, the central government and local administrations made a major effort to cope with the health emergency in towns. Italy adopted a code of urban hygiene and established an elaborate network of health offices to support local administrations'

The transition to modernity

Figure 2.3

Naples gulf view. Public domain photo from Pixabay.

sanitisation actions. This was done in the wake of the Sanitisation Act for Naples (Risanamento), approved in 1885 after the cholera epidemic, and the Public Health Code (Codice sulla salute pubblica) of 1888, later improved with further sanitary legislation in the 1930s.

A major turning point in the transformation of city environments was the introduction of innovations in modes of access to water and the drainage of outflows. A first leap forward was made between 1888 and 1899, with the building of a network of aqueducts serving forty per cent of Italian towns. Still, as late as the 1930s only 10,000,000 inhabitants (out of 40,000,000) were served by aqueducts. In 1975, towns reached by an aqueduct were 6,835 out of a total of 8,035. The spread of water infrastructure did not occur homogeneously across Italy, but reflected the traditional gap between the North-Centre and the South. For a long time, the situation of aqueducts and sewage was distinctly worse in the South and the islands than in northern and central Italy. It partly improved after World War II thanks to a general plan for aqueducts approved in

1963. Nevertheless, as late as the end of the 1990s, over half the population of the Italian South complained about water shortages during long periods of the year.

In urban areas in north-central Italy, the 'town company' model – the so-called *municipalizzata* – created to establish large service infrastructures, took hold more than it did in the South, and as early as the beginning of the twentieth century. Water management is one of the sectors where *municipalizzate* became more widespread, along with gas and electricity. The new infrastructures were town property, or built at their own expense by private subjects who retained management rights. Many were built by foreign companies, such as the French Société Générale des Eaux and Compagnie Générale des Conduites d'Eaux.

The water incorporation process mobilised a broad and complex range of resources and ecosystemic equilibria, both within and outside towns. In the first stage of this process, aqueducts bringing water from far away were planned in many middle-to-large towns. Eventually, however, the most frequently used method became drilling down into deep reservoirs or drawing water from rivers. In Milan, in spite of several plans to build aqueducts and pipes bringing in water from the Val Brembana, the choice eventually made was to drill deep wells right under the city, at much lower costs. Turin drew its water from the Sangone torrent through a circa ten-kilometre pipe, supplemented by intake apparatuses in Venaria. In Ravenna, a plan to build a fifty-kilometre-long consortium aqueduct bringing water from Mount Fumaiolo met with failure. The town ended up drawing its water, instead, from the Marecchia river aquifer in Torre Pedrera di Rimini.

Entire ecosystems were altered by the lowering of aquifer and riverbed levels. New outlets were found for the increasing quantities of wastewater from homes and industrial plants in natural underground cavities, rivers and the sea, causing serious pollution problems. Some aqueducts drew their water from springs. For example, Genoa drew water from the Bisagno and Lavena torrents. In Rome, the springs of the Val d'Arsioli at Monte di Tivoli were used, and the ancient Marcian aqueduct was restored. Another ancient aqueduct was restored in

Naples, that of Serino, fed by the springs of Acquaro-Pelosi and the Polla di Urcioli.

The aqueduct and water drainage infrastructure was reinforced under Fascism in the framework of the land reclamation policy implemented by the regime to address Italy's swamp and drought problems. This led to the construction of the Puglia aqueduct, decreed by Act 1365 of 1920. When it first became operative, this imposing work comprised 244 kilometres of main canals, fed by the sources of the Caposele. It was expanded after World War II by establishing reservoirs fed by the springs of Occhitto, Locone, San Giuliano, Pertusillo, and Sinni. In the early 1930s, the Monferrato aqueduct was built, fed by the Dora Baltea river. Further noteworthy aqueducts were those of Simbrivio, bringing water from Apennine springs to the Provinces of Rome and Frosinone, and Ruzzo, which served many towns in the Province of Teramo.

The latter was a major work, but also a conflicted one. There were protests and mobilisations against what was seen as a misappropriation of the water resources of the prevalently farming districts along the path of the aqueduct's pipes and canals. But the real problem lay in the environmental incompatibility between agricultural, civil and industrial uses of water. Water was also employed as a place to dump waste, as a raw material in industry, and in hydroelectric plants, which constituted the main source of energy in Italy, a country that is scarce in coal.

A worse problem than competition for water resources was the dumping of industrial waste in rivers, wells and aquifers, making their waters unfit for domestic and agricultural use. Paradoxically, the modernisation of the water supply and sewage infrastructure, while it radically improved living conditions in towns, had severe repercussions for the environment. The cycle that allowed organic waste to be re-employed in agriculture, restoring part of the consumed resources, was irreversibly broken. On the one hand, agriculture lost its self-sustaining source of natural fertilisers, which were replaced by chemical ones. On the other, aqueducts and sewage systems led to an unprecedented consumption of water for domestic and industrial uses, mobilising an immensely greater quantity of water than in the past.

New water conveying methods were needed, as well as areas to dispose of solid and liquid waste; and new forms of water pollution needed to be dealt with. To this day, we are still paying a price for wrong solutions to these problems, which did not take account of long-term environmental impact and the need for the reproduction of natural resources. In general, modern water supply and drainage technologies have played a crucial role in increasing pollution levels while improving public health in cities and their neighbouring areas. The water infrastructure was often built with no regard for its side-effects on the surrounding areas. Those concerned with protecting the health of the urban population and freeing cities from pollution overlooked a whole range of problems precisely because they ignored the environmental consequences of the use of water. This approach – the same that had inspired the legislation on insalubrious industries in several countries in northwestern Europe ever since the early 1800s – was adopted in Italy in the sanitary legislation of 1888, and subsequently reiterated in the Fascist legislation of the 1930s and then in the pollution act of 1966. Only after World War II did technicians and scientists start to look at water as a scarce and vital resource for human health and the environment. Nationwide legislation adopting this perspective did not come until 1976, when Act 319 – the so-called 'Merli Act' – was approved, aimed at protecting water from all forms of pollution and including measures to guarantee its replenishment.

Town planners tried to address the problems generated by Italy's impetuous and chaotic urban growth. This was the time when the first general town plans were drawn up: Paris in 1848, Vienna in 1857 and Barcelona in 1859. In Italy, the first city to draw up a town plan was Florence in 1864. Sanitary considerations required a redefinition of the urban space, new infrastructure and new transportation facilities to connect residential and industrial areas. Social values replaced aesthetic ones in the planning of urban green. This replacement was not a result of town policies alone. The garden city movement, founded in England by Ebenezer Howard in the late nineteenth century and promoted and endorsed by associations sprung up in middle-class milieus, had spread to several European countries. It promoted an urban model that strove to offer a third alternative between town and country by building 'garden cities' that would allow for a full-ness of social life while doing away with the typical shortcomings of urban conglomerates. In Italy, this trend went hand in hand with the struggle of doctors and hygienists against diseases such as tuberculo-sis. 'Garden cities' were built in several Italian towns, including one along the Via Nomentana in the suburbs of Rome.

In the decades between the nineteenth and the twentieth century, Europe underwent a radical change in its relationship with nature. This change in culture and perception actually had its roots back in the eighteenth century, but had reached its full bloom in the course of the nineteenth. Sea, mountains, woods and water bodies were no longer perceived under their most terrifying guises; rather, they became a source of pleasure, beauty, relaxation and comfort. People's relationship with their own bodies also changed. Gymnastics became increasingly popular, as did the quest for open-air spaces and new dress styles adapted to these new pursuits. Human beings' relationship with animals altered too, with more and more people keeping dogs or cats as pets.

Water, in particular, acquired a new centrality in medicine during the nineteenth century through the spread of hydrotherapy. The therapeutic properties of thermal water and sea bathing were advertised broadly in the press. Eventually, people started going to the sea-side for leisure as well as health reasons. The 'beach fever' first took hold among aristocratic families of northern European countries in the second half of the eighteenth century. By the nineteenth century it had also spread to other social classes. The Italian peninsula had long been the favorite destination of Grand Tour travellers. This tradition paved the way for the economic exploitation of the country's environmental beauties. A number of areas founded their wealth and fame on their thermal springs, including Ischia, Montecatini, Recoaro, San Pellegrino, Salsomaggiore and many others. The increasing popularity of sea bathing, in turn, favoured the rise of tourist resorts along the Italian coast. The first areas where sea bathing spread were the coast of Liguria, the area around Viareggio, the Lido di Venezia, the upper Adriatic and coastal cities like Naples and Palermo and their surroundings. In the course of the twentieth century, tourism became accessible to new social classes; by the 1960s it had become a mass phenomenon. Driven by small and medium-sized companies and the very active engagement of local associations and administrators, today tourism is one of the leading sectors in the Italian economy. However, while until the second postwar period its exploitation of natural resources remained virtuous, with weak negative effects on the environment, today it threatens to become a deterioration factor, especially due to the way in which seaside resorts are expanding.

3. The impact of early industrialisation

Ever since the origin of modern growth, Italy leaned towards an industrial development inspired by the philosophy of laissez-faire, essentially unregulated, with its sights set on profit and indifferent to pollution issues. The impact of industrialisation on the environmental balance increased from the last decades of the nineteenth century, when industrial production peaked, particularly between 1896 and 1908. The localisation of industries was based on economic and infrastructural considerations, such as proximity to roads, railway stations or ports, or ease in recruiting labour. Industrial development hence mainly occurred in and around the larger towns. It was initially concentrated in Liguria, Lombardy and Piemonte, the so-called 'industrial triangle', but subsequently spread to northeastern and central Italy and parts of the South, particularly Campania. According to estimates provided by Vera Zamagni, in 1911 55 per cent of industrial added value was produced in the industrial triangle, 29 per cent in the Northeast and Centre, and sixteen per cent in the South. While small farms and workshops continued to operate, soon more modern and technologically advanced sectors arose, including metalworking, electricity, shipbuilding, concrete production and the chemical and automobile industries. The international crisis of 1929 was followed by a remarkable recovery of Italian industry, partly thanks to the foundation of the Iri – Istituto di ricostruzione industriale – whereby the State ended up controlling most Italian industry, including ironworks, shipbuilding, electricity, train production etc.

Like the urban environmental problems discussed above, the

environmental problems generated by industrialisation were construed and addressed as being essentially hygienic and sanitary in character – that is, exclusively from the perspective of their repercussions on human health. An act approved by the Crispi government in 1888 as part of a broader reform of public healthcare introduced the concept of insalubrious industries and regulated their activities. Dangerous factories or factories producing harmful emissions were to be included in a special list and divided into two classes. The first comprised those that had to be removed to isolated locations in the countryside, the second those requiring special precautions for the safety of the neighbourhood. The list, published in 1895, included a rather broad range of industries, from those emitting hydrochloric or sulphuric acid or processing arsenic, lead or chrome to those producing animal waste or manipulating putrescent materials such as hemp and linen. The second class included wax and paper factories and ironworks, which employed less polluting raw materials.

At the beginning of the twentieth century, over 20.7 per cent of the 97,000 industries in Italy (including construction industries) were classified as insalubrious. More than 37 per cent of these were located within the industrial triangle. Here industrialisation had concentrated around urban centres, both large towns like Turin, Milan and Genoa and many medium-sized ones like Savona, La Spezia, Novara, Brescia, Bergamo, Como and Monza. The rest of the harmful plants were located in suburban districts that had already been involved in proto-industrial production, such as the Valle dell'Olona with its textile industries, the Valle dell'Agno and the area around Schio, the coasts of Venezia Giulia and Emilia, the Val Bisenzio in Tuscany, the Terni valley in Umbria and Campania with its textile and food industries.

The hygienist approach focused only on the protection of people's health while ignoring the effects on nature. This underestimation of damage and risk was partly a result of limited technical and scientific knowledge about the noxiousness of both gaseous and liquid emissions. There was a reliance on the self-purifying properties of air and the diluting properties of water. Building a wall or a chimney stack, or dumping residues underground or in a well was regarded as sufficient protection against dangerous waste or emissions. And this is not to mention the notion that acid steam and coal smoke had beneficial effects. The inadequacies of the Act of 1888 and of all those that followed in its wake, however, did not merely reside in the hygienist principles it was founded on. The problem with this legislation is that it addressed the effects, not the causes. For example, as Simone Neri Serneri explains, an entrepreneur was free to choose where to locate his industry, and only subsequently was required to take measures to avoid its harmful effects. Besides, the application of the law was entrusted to an untrained administrative apparatus, oscillating between the defence of the interests of industry and the defence of citizens' health. With some significant exceptions, especially in the big cities, local authorities did not have the knowledge, the tools or the financial strength to stand up against the interests of industrialists, when they were not on the industrialists' side to begin with. Town administrations had several duties, including: classifying industries in the first or the second class; establishing rules; specifying minimum distances of factories from housing; ascertaining if industries in the first class had adopted the prescribed measures, or if those in the second class had adopted the special precautions required for them to be allowed to remain in residential neighborhoods; and sanctioning 'insalubrious' manufacturing activities. If a legal dispute arose, the procedures were elaborate and convoluted; indeed, they often required so much time that they actually granted impunity to transgressors. In spite of this, controversies and conflicts did arise, involving not just citizens affected by noxious factories located in their towns, but also farmers and tourist resort operators. The advance of urbanisation and industrialisation had taken water resources away from traditional local economies by building aqueducts bringing water to towns from far away, modifying water management balances in various ways or polluting water and thus making it unfit for domestic and nutritional use, or for farming or fishing. The environmental conflict was prevalently a conflict between incompatible uses of natural resources, between urban and industrial uses, but especially between industrial and agricultural uses. Pier Paolo Poggio has spoken of 'farmer resistance'; rather than being a clash between the forces of progress and those of conservation, this phenomenon reflects an epochal change in the use of natural resources and in environmental equilibria, a change that lies at the origin of contemporary society and its environmental issues.

The environmental history of industry is especially the history of the deep changes undergone by areas where new factories were established. Entire ecosystems faced the loss of earlier productive uses, the expansion of factory buildings and infrastructure, depletion of resources, and contamination of the air, ground and water. Notable cases include the sugar factories of the Po River valley, the paper mills along the slopes of the Apennines and the Acna chemical plant at Cengio, which during the early decades of the 1900s dumped its processing waste in the Bormida river, polluting it for dozens of kilometres. The noxious exhalations and waste of the chemical industry located in the town of Rumianca in the early twentieth century affected several towns, all the way to Lake Maggiore. Further notorious polluters include the Caffaro factory in Brescia, the chemical industry of the Sacco river valley in the Province of Frosinone, the Terni valley industries, Porto Marghera in Venice and Bagnoli in Naples, a suburban area where an iron, chemical and asbestos factory was built at the beginning of the century.

The sanitary legislation of 1888 and the regulations issued to enact it, which were taken up again in the sanitary acts of 1934, allowed industrial wastewater to be drained into watercourses within cities if adequately purified and if certain special precautions had been taken. Drainage of wastewater was only allowed in channelled watercourses with waterproofed banks. In spite of its weakness, this sanitary legislation did produce some effects. Besides ousting from residential areas some very noxious traditional industries, such as tanning or the production of fertiliser from animal residues, it encouraged the building of new plants outside of town centers, in periurban areas or metropolitan hinterlands. The isolation of industries from cities found full implementation in the 1930s with the institution of the 'new industrial zones', which included Venice, Livorno, Bolzano, Ferrara, Apuania, Palermo and Rome. In some cases, for example those of Porto Marghera and Carrara, these areas eventually grew into major hubs of the chemical, mechanical or metallurgical indus-

Figure 2.4

Vajont dam, northern Italy. Public domain image from Wikipedia.

The transition to modernity

Figure 2.5

Hydroelectric power plant, Lombardy. Public domain image from Wikipedia.

try. The relocation of factories outside urban centres encouraged an even more destructive use of natural resources, especially as regards atmospheric emissions and the dumping of liquid and solid waste in water and in the soil. As Neri Serneri has shown, the creation of intense industrialisation zones, which brought to extremes the principle that industries with their noxious effects should be separated from the city, actually resulted in further loss of control of environmental impact under legislation that was already weak to begin with, leaving industries ample freedom to pollute.

Urbanisation and industrialisation caused not only a multiplication of sources of pollution of water resources, but also an unprecedented growth of the mobilisation of water. The largest quantities of water were used to power hydroelectric plants. Italy is scarce in fossil coal resources, which it used to import from France and England. Therefore, starting from the beginning of the twentieth century, it fuelled its industrial development with hydroelectric energy. The country's mountainous nature and abundance of *carbone bianco* ('white coal'), as its water resources were called, facilitated its exploitation. For a long time, hydroelectricity accounted for a large percentage of all the electricity produced in the country. In 1905, seventy per cent of utilised power was of hydroelectric origin. In 1925, Italy had 588 hydroelectric plants, which grew to 833 in 1938.

A fundamental role in Italy's hydroelectric energy programme was played by politician and *meridionalista* ('southernist') Francesco Saverio Nitti, who was minister of Agriculture, Industry and Commerce from 1911 to 1914, and Prime Minister from 1919 to 1921. Nitti regarded hydroelectric power as the primary resource that would allow the Italian South to rise anew and kick-start its industrial development. The objective was to produce hydroelectric power while simultaneously addressing hydraulic threats in the mountains – made worse by deforestation. The combined exploitation of water for hydroelectricity and irrigation led to the disappearance of latifundia. Nitti's idea was to employ water formerly constituting a hydrogeological threat or a hotbed of diseases, such as malaria, for good uses – to produce energy or for irrigation. To achieve this, mountain slopes needed to be reforested and stabilised. His vision was to inspire the concept of integral land reclamation, formulated during the Fascist period and fully implemented in the second postwar period.

In 1904, Nitti had a special act for Naples approved, which favoured industrial growth in the city, but met with much resistance insofar as it provided for the establishment of autonomous management for the Volturno and Tusciano rivers. During the Giolitti administration, various acts and special legislation were issued to address the environmental impact of industrialisation. They included the act for Basilicata of 1904, that for Calabria in 1906 and another of 1907 instituting a magistrate for the management of the Po river basin. Another act approved in 1912 was meant to facilitate the building of artificial lakes in southern Italy. In 1913, a series of artificial lakes was planned in the Tirso area in Sardinia and the Neto area on the Sila plateau; they were completed after the war. The approach to water management had thus reached another turning point. During the 'hygienist period', it had been perceived as the resource that would solve the issue of the degradation, at once environmental and moral, in which urban populations lived; now the technocratic mentality regarded it as the principal resource for national economic development.

4. The protection of nature

To detect the first traces of a modern environmentalist awareness in the Western world, we need to go back to the time when the so-called 'bucolic ecology' tried to restore the link between human beings and nature after it had been severed by the scientific revolution, which had regarded the natural world as separate from human beings and ordered as a machine. The English naturalist Gilbert White was an exponent of this trend. In his *The Natural History of Selbourne* (1789) he reported the observations he had made for decades on the animals in his village, conveying a sentiment of respect for nature and a nostalgic longing for pastoral harmony between human beings and the environment.

Among scientists, the first significant contribution to the construction of an ecological thought came from Linnaeus, the Swedish botanist Carl von Linné. In his *Oeconomia naturae* (1749), he offers a portrait of nature as an endless process characterised by complex geobiological interconnections. In his *The Origin of Species* (1859), Charles Darwin gave an in-depth critique of mechanism and developed an organic conception of nature. His evolutionist theories postulated a close relationship between human beings and other living species, and between living species and the physical world.

In Italy, an awareness of the theme of protection of nature first arose in the context of the idealist revival of the late nineteenth century. The reaction to positivism and the dominion of science gave rise to a whole new interest in monuments and the landscape. But this new emphasis on the defence of heritage was also rooted in the idea that the construction of a united nation should go hand in hand with a new generalised awareness of its natural and artistic beauties.

New reflections on the relationship between human beings and nature can already be found in earlier periods. For example, the abbot Antonio Stoppani's *Il Bel Paese. Conversazioni sulle bellezze naturali, la geologia e la geografia fisica d'Italia*, published in 1876, enjoyed great popularity. The book allowed its readers to become acquainted with Italy's landscapes and nature spots. Translations into Italian of a number of foreign books were also influential. These books included Alexander von Humboldt's *Kosmos*, published in Italy in 1846; George Perkins Marsh's work *Man and Nature, or Physical Geography as Modified by Human Action*, translated in 1870; and the writings of the evolutionist scientist Ernst Haeckel, who coined the term 'Ökologie' to define the field studying the relationship between living organisms and their environment.

A strong protectionist sensibility arose in Italy between the last decade of the nineteenth century and the Great War. This was a unique period in the country's history, at least until the rise of political environmentalism in the last decades of the twentieth century. It witnessed a profusion of conferences, lectures, exhibitions, public initiatives in defence of artistic monuments or places of special landscape or naturalistic value, various kinds of publications, journals, information and teaching activities, etc. This Italian environmentalist movement wove a close-knit web of contacts and exchanges with nature protection movements in other countries. The National Trust for Places of Historical Interest or Natural Beauty was founded in Great Britain in 1895, where it had been preceded by the Commons, Open Spaces and Footpaths Preservation Society, created as early as 1865. The Société pour la protection des paysages was founded in France in 1901 and the Heimatschutzbund in Germany in 1904; only two years later, more than 100,000 citizens were involved in the latter's activities. There was a growing awareness that the protection of nature required action and measures transcending national boundaries. The most significant moment in the cooperation between these environmentalist movements was the Berne conference of 1913, whose participants committed themselves to engaging in large-scale information campaigns to stop the destruction of the plant and animal species most threatened by international trade.

In that same year 1913, the Lega per la Protezione dei Monumenti Naturali and the Comitato Nazionale per la Difesa del Paesaggio e dei Monumenti were founded in Italy – two different institutions, but with largely overlapping objectives. Through both, the different facets of proto-environmentalism managed to promote action in defence of *monumenti naturali* or, as they were later called to put the accent on aesthetics, *bellezze naturali* ('natural beauties').

This early environmentalism, well described by Luigi Piccioni in his book *Il volto amato della patria*, was not a homogeneous movement, as it encompassed different notions of environmental protection. One, held by natural scientists, was more sensitive to the destructive impact of modernisation on natural resources and ecosystemic equilibria. The Società Botanica Italiana (1888) and Società Zoologica Italiana (1900) struggled to preserve Italy's fauna and flora by combatting the gathering of rare plant species and the sale of feathers and other bird parts, hides and wild animals. The institute for the study of sea life founded in Naples by Anton Dohrn in 1873 also played an important role in this field. The association Pro Montibus et Silvis (1897), which published a journal called *L'Alpe*, addressed issues of forest management and hydrogeological risk. Although it was in the minority, the scientific component of environmentalism played a major role in assisting in the institution of Italy's first natural parks.

In another component of the environmentalist movement, the notion of protection of nature tended to combine the appreciation of natural and cultural treasures with an outdoors lifestyle and regenerating sport practices. In 1863, the Italian Alpine club was founded, which combined a love for sports with research and teaching. With the creation of the Touring Club Ciclistico (Bicycle Touring Club, later Touring Club Italiano) by Luigi Vittorio Bertarelli, the tourist and sports-oriented component of the nature protection movement entered a phase of great élan. By the end of the 1910s, the Touring Club Italiano already had as many as 70,000 members. By means of maps and magazines, technical reports and guidebooks, these two associations played a seminal role in spreading knowledge of Italy's natural and cultural treasures, thereby contributing to strengthening a sense of nationhood in the newly united country, especially among the ruling and middle classes. Down to this day, the 'red guidebooks' of the Touring Club constitute a major source for popular knowledge of Italian artistic and landscape heritage.

The third component revolved around a group of politicians and leading public officials who engaged in a number of institutional battles in defence of the environment. This component was prevalently animated by an aesthetic and patriotic conception of the protection of nature. It found its most eloquent expression in the Direzione per le Antichità e Belle Arti (Direction for Antiquities and the

Fine Arts) of the ministry of public education.

This institutional activity led to Act 778 of 1922, entitled 'Per la tutela delle bellezze naturali e degli immobili di particolare interesse storico' (For the protection of natural beauty spots and buildings of special historic interest). The act was the outcome of a long process that had started with Act 185 of 1902 – complemented by another act issued in 1903 – for the conservation of monuments, works of art and archaeological artifacts. In 1909, another similar act had been approved. It was much influenced by a French act of 1906 for the protection of monuments and natural sites of artistic interest, subsequently complemented by an act of 23 June 1912 that extended this protection to villas, parks and gardens. Interestingly, a clause allowing expropriation for public utility was edited out of the final text of the act of 1922. The difficulty of conciliating private property with the collective function of ecosystems was to be a leitmotiv in environmental protection policies.

Unlike earlier legislation, however, the act of 1922 for the first time placed a broad range of heritage under protection. It established the principle that any modification of privately owned nature spots needed to be authorised by the Ministry. It also forbade the damaging of the landscape or the undermining of enjoyment of nature spots as an integral whole – that is, as scenic views. These principles were reconfirmed in Act 1497 of 1939, entitled 'Protection of natural beauties'.

In spite of the approval of this act, the Fascist period marked a weakening of the protectionist movement and a decline of interest in environmental themes. After the war, the protection of landscape was given legal sanction in Subsection 2 of Article 9 of the Constitution, although it was not given as much importance as in broader and more detailed proposals put forward during the constitutional debate. In his book *La distruzione della natura in Italia* (1975), Antonio Cederna observes that, unlike the constitutions of other European nations, Italy's overlooks themes such as environmental security, the conservation of nature and the environmental and social repercussions of urbanism. The constituents' lack of interest in environmental themes, he adds, was to have serious consequences for the history of republican Italy.

It is true, however, that juridical debate and the jurisprudence of the Constitutional Court very slowly spawned a broader and more global conception of the environment. The reform of Chapter Five in 2001 espoused this conception, referring not only to Article 9 but also to articles relative to the inviolable rights of man (Art. 2); the equal social dignity of all citizens and the Republic's task of removing all obstacles to the full development of the human person (Art. 3); and the protection of health as a fundamental right of individuals and an interest of the collectivity (Art. 32).

Concomitantly with the drafting of this legislation between 1921 and 1922, a catalogue of natural attractions was drawn up. A fill-out form listing a number of categories of natural attraction was distributed to associations and citizens. Interestingly, the categories they selected somehow reflected the different trends within protectionism. They ranged from natural beauty spots to sites of historic or literary relevance, from woods, villas, parks and gardens to rare animal species, and even rare folk costumes. The categories of 'natural monument' and 'natural beauty' prevailed over those of scientific, historic or literary relevance.

From a political standpoint, the first nature protection movement in Italy was rooted in the liberal ideas of Giovanni Giolitti and Francesco Saverio Nitti. Remarkably, the theme of the defence of the environment found most attention in the context of the technocratic ideas promoted by Nitti, which still strongly emphasised the need for economic growth. For Nitti, it was a matter of taking action that would simultaneously address environmental threats and allow the use of water as an energy source, meeting the needs of industry while solving the environmental issues that had long been plaguing Italy, and especially its South. Economic growth and the improvement of environmental conditions should go hand in hand; indeed, one was the prerequisite of the other. Nitti's relationship with protectionism was thus complex. His technocratic approach met with opposition locally and from protectionist associations such as the Club Alpino Italiano and the Touring Club, due to the environmental destruction it involved – digging artificial canals, cutting down trees and altering the beauty of landscapes. On the other hand, Nitti gave protectionism strong moral and substantial support in opposing hunting and promoting the creation of the Abruzzo National Park.

Apart from some personal stances, environmental issues lay essentially outside the sphere of interest of other political cultures. In this regard, Italy differs from Germany, which saw the rise of a movement for the protection of nature with a strong emphasis on cultural identity founded on people's belonging to a community and an area, in the broader framework of a ruralist ideology exalting *Volk*.

Besides the act of 1922, the most important achievement of the Italian nature protection movement was the institution of the national parks of Abruzzo and Gran Paradiso (1922–23), as the result of action taken at the beginning of the century by the Pro Montibus committee for parks. Although supported by all the currents in Italian environmentalism, the establishment of the two parks was fraught with difficulties. The choice was made from a broad selection of areas, including the Livigno valley on the border with Switzerland, the Phlegraean Fields near Naples, the Abetone and Teso forests and Argentera range on the Maritime Alps, the Gennargentu and the Sila range. Piccioni argues that the choice was influenced by the fact that the area of the two parks had been royal game reserves and their exceptional natural heritage had hence been effectively preserved. The choice of the upper valley of the Sangro river for the Parco Nazionale d'Abruzzo had been preceded by a decree of 1913 – subsequently converted into an act – for the protection of the Abruzzo chamois, a species that was at high risk of extinction due to aggressive hunting. Nitti, who at the time was minister of agriculture, was one of the endorsers of the decree. Later on, in 1934, the National Park of Circeo was instituted, and a year later the National Park of the Stelvio.

Italian protectionism had long ago started discussing the possibility of protection extending beyond individual spots or monuments to embrace much larger and more complex areas. Through travels in the United States by Italian scholars and naturalists, the Italian scientific community had come into contact with a culture that had started to create natural reserves early on. The Yellowstone Park, the first of its kind in the world, had been instituted as early as 1872. The American model, however, based on the concept of 'wilderness', had

The transition to modernity

the advantage of vast uncontaminated reaches, even though threatened by the advance of modern civilisation and mechanised farming. Europe's long history of settlement and high demographic density, as well as the deep and complex changes vast areas of its farmland had undergone, called for entirely different solutions. And this was especially true of the Italian peninsula.

CHAPTER THREE.

ENVIRONMENT IN REPUBLICAN ITALY

1. The environmental implications of an energivorous society

In Italy, the energy revolution founded on the transition from renewable to non-renewable sources began to unfold in the late nineteenth and early twentieth centuries. It was only after World War II, however, that it brought about the rise of a social and economic organisation multiplying the quantity of employed energy to an unprecedented scale – that is to say, the transformation of Italy into a major industrial power. This transition was not sudden. Coal replaced wood during the first fifty years after unification. Petroleum replaced coal in the decades after World War II. According to data provided by Paolo Malanima, between 1869 and 1913 the percentage of energy from wood decreased from 51.4 to 21.2 per cent, while that of energy from coal grew from 7.5 to forty per cent. Between the 1950s and 1970s, energy from petroleum, which had become available thanks to imports from the Middle East, grew from 18.4 to 72 per cent, while energy from coal decreased from 27 to seven per cent. An extraordinary growth began in all sectors of the country's productive and everyday life, including industrial activities, urban areas, transportation, infrastructure, agriculture and families' consumption models. This epochal transition marked the beginning of a period of widespread affluence characterised by increasing income and consumption, an increase in the average life span and the disappearance of typical diseases of traditional societies such as typhus, cholera and malaria.

This exceptional improvement of life conditions, however, came at a cost. The advent of energivorous society determined very strong pressure by human activities on the environment and rapid dissipation of natural resources, both renewable and non-renewable. This pressure produced a change in the character of environmental risk and pollution, and expanded its effects immeasurably, even very far from its places of origin. A new historical phase had begun, in which the environmental question transcended national borders to acquire, in recent decades, a planet-wide dimension through global warming and consequent climate changes, the expansion of the ecological footprint of wealthy countries and international traffic in waste.

Italy's exceptional industrialisation went hand in hand with urban expansion, a growth of sanitary and transportation infrastructure and electrification. Towns and industries incorporated and expelled increasing quantities of nature, in the form of water, soil and energy, on the one hand, and emissions and waste, on the other. This had a major impact on the ecosystemic equilibria of vast areas. The industrial sector absorbed most energy, peaking at 49.2 per cent in 1965, to then decrease, at first slightly, then more significantly after the 1980s. The household sector, which used energy for heating, cooking and lighting, absorbed about thirty per cent. This figure was rather stationary, but reached a peak of 34.5 per cent in 1975. Throughout the first three postwar decades, energy absorption in the transportation sector hovered around twenty per cent, until the 1980s, when it started growing with the advent of the 'age of the automobile'.

Between the 1950s and 1960s, demographic growth was accompanied not only by a move towards metropolitan areas, but also by a general reshuffling of the population within the Italian peninsula. About fifteen million people changed their town of residence. People moved not only from South to North, but also from the southern ranges of the Alps to the Venetian and Tuscan littorals, and to Milan, Turin, Genoa, Venice and Bologna. Within southern Italy, there was a move from the slopes of the Apennine to areas along the coast and in the plains reclaimed through public works financed by the Cassa per il Mezzogiorno, a major state fund for the South.

Between the 1950s and 1960s, industrial production grew by 180 per cent and energy consumption by 134 per cent. This growth was fed by the expansion of large and middle-sized businesses in the Northwest, scattered or clustered industries in the Northeast, and some large industrial poles in some areas of the Italian South. New urban-industrial areas were formed or existing ones expanded, including Sesto San Giovanni, Mestre-Marghera, Piombino, Terni, Bagnoli, Manfredonia, Brindisi, Taranto, Augusta-Priolo, Gela and

Porto Torres. For decades, these areas had a very destructive impact on environmental balances, and some still do today.

The chemical, petrochemical and iron industries were the sectors that absorbed the most resources – prevalently water – and energy. Pollution was worst in the surface and underground waters of the urban areas of Milan, Turin and Genoa, as well as in the river, sea and lake waters where the large urban and industrial areas dumped their waste. The affected areas included the coasts of Veneto and Romagna; those of Liguria and Tuscany; vast areas of the Roman and Neapolitan conurbations; the coast of Puglia near Bari, Taranto and Brindisi; the coasts of Sicily and Calabria along the Straits of Messina; and the coast of Sicily between Catania and Siracusa, and near Palermo and Marsala. As to air pollution, its distribution largely matched that of water pollution. It mainly affected the large conurbations of the North, the metropolitan areas of Rome and Naples and long stretches of the littorals of Liguria, Veneto, Puglia and Sicily. Early on, growing mortality and disease among workers, especially in or near large or middle-sized towns, revealed which energy production and manufacturing sectors were the most polluting, albeit only very approximately. The main factors in air pollution were sulphur dioxide, nitrogen oxides, hydrocarbons and particles. Water was polluted by scarcely biodegradable toxic residues of industrial processes. Industrial waste – which over the last decades has become one of the country's most serious environmental issues – was often dumped into unauthorised landfills or in watercourses. It included iron scraps, sludge, ash, bulky items, chemicals, plastic etc.

The countryside also became increasingly industrialised. It was thus impacted on by new, powerful sources of fossil energy as a result of the spread of chemical products – fertilisers, pesticides, weed-killers – and machines, which boosted agricultural productivity at extraordinary rates. This trend can be traced all the way back to the Fascist period, when the state's involvement in the economy had been especially intensive. In the mid 1920s, the issue of land fertility had been the main theme in the regime's *battaglia del grano* ('Battle for Grain'), a propaganda campaign aimed at making grain agriculture the driver of Italy's economic growth. In addressing the masses,

Mussolini had repeatedly stressed that one of the purposes of the Fascist regime was to increase the 'fecundity of the soil'.

The highest increase in the use of chemical fertilisers, however, was in the 1950s. Specifically, between 1947–48 and 1958–59 the use of nitrogenous fertilisers increased by 334 per cent and that of sulphur dioxide by 224 per cent. Agricultural production grew accordingly. Corn yields, for example, went up from 2.84 tons per hectares in 1956–58 to 7.2 in 1980–82, while the overall cultivated surface decreased markedly. Sugar beet yields grew from 30.48 to 48.05 tons per hectare. Tomato crops doubled their yield. As to tree crops, the yields of most increased, particularly those of citrus, and even more those of apples, which rose from 12.95 to 30.56 tons per hectare, and peaches, 70.2 to 174.4.

The growth of the productivity of labour was no less astounding. The spread of new technology determined a drastic fall of the farming population – from 8,261,000 units in 1951 to 4,023,000 in 1969 and less than 3,000,000 in the early 1980s. A massive growth of mechanisation and motorisation thus went hand in hand with an exodus from the countryside. In terms of productivity, the introduction of machines and motors more than made up for this shrinking of labour.

The development of agriculture was promoted by a series of policy actions granting credit and benefits, notably two quinquennial plans in support of agriculture implemented in the 1960s, the so-called 'Green Plans' (*Piani verdi*). A national agricultural plan approved in December 1977 increased public subsidies. Such national measures were supplemented by European Community policies aimed at reducing fluctuations of income from agriculture and protecting the food market to sustain the food sources of Europe's population. These policies had significant environmental implications, and not always positive ones. They eventually resulted in making agricultural production dependent on an artificially created demand, to the sole end of granting a profit to whole sectors of the economy. The same phenomenon occurred in other European countries, such as Great Britain, France and Germany.

As Piero Bevilacqua has explained, the postwar success of industrial agriculture, in Italy as well as in the other main European countries, was the result of an 'artificialisation' of the biological cycle, which increased its dependency on fossil energy without actually increasing fertility in the sense of overall potential for production. This process was aggravated by several factors, including loss of organic substances and a consequent worsening of the physical structure of the soil in many intensive cultivation areas, weakening of crop resistance to pathogens as a result of increasing use of fertilisers and pollution of water by chemical products administered in doses exceeding crops' absorption capabilities.

The industrialisation of the countryside also brought on a radical transformation in animal husbandry, involving its separation from agriculture and the spread of forced stabling and cooping, with serious consequences such as loss of natural fertility and the rise of diseases due to poor-quality feed. This way of raising livestock is not aimed at producing better and healthier food, and often depends on an artificially created demand.

Today, agriculture has become the sector consuming the largest quantity of water resources. According to data relative to 1999, residential and industrial uses of water resources accounted for only forty per cent of the total (about twenty per cent per sector), while irrigation accounted for 49 per cent. The remaining eleven per cent was used to produce energy.

This transformation affected all of Italian agriculture. Until the 1950s, technological innovations had not affected traditional farming work and management systems. In the following decades, instead, technology became the main factor in the organisation of production. The new machines shaped the agrarian environment. New crop layouts were adopted to allow the passage of forage harvesters, self-loading carts, olive harvesters and elevating platform trucks for fruit picking. Agriculture moved away from rugged places where machines could not move about.

The absence of human beings and the spread of machinery powered by new mighty energy sources has had a tremendous impact on agrarian systems. In just a few decades, it has radically redesigned a farmed landscape which had characterised the Italian countryside for centuries, in spite of significant changes. In the Po River plain, the dense green grid of rows of trees and vines has been replaced by specialised crop fields. In central Italy, mono- and bicultural crop-ping has replaced mixed cultivation. In the South, irrigation and land reclamation works, largely financed by state subsidies, have allowed a transition to an irrigated agriculture and favoured the descent of arboreal crops from the hills to the plain.

Machines' replacement of men has accelerated the decline of vast stretches of once productive land. In mountain and high-hill ar-eas – the most affected by the agricultural exodus – the raison-d'être of labour-intensive cultivation adapted to sloping terrain was no longer there. Their unsuitability for mechanised farming pushed vast areas into economic marginality. Rural emigration caused a decline of land maintenance and drainage work, with a consequent worsen-ing of already serious erosion and rainwash.

Between the 1950s and the 1970s, the Italian economy grew without adequate consideration of the destruction its growth was wreaking on ecosystemic balances, and without assessing the costs and 'environmental debt' it was bequeathing to future generations. While public authorities were sensitive to the issues of soil defence, reforestation of mountain slopes and reclamation of land in the plains, and took policy actions to address them, the same cannot be said for water and air pollution or the dumping of waste and toxic substances. The regulation of the outflows of urban-industrial ecosystems was implemented belatedly and was not always effective in limiting dam-age to ecosystems and human health. The interests of public and pri-vate industry strongly influenced governmental and legislative action. The dominant paradigm of the time was that of growth and a general faith in the benefits of a development model founded on industrial production and expansion of the consumer base.

Public institutions' reaction to pollution issues was slow, uncer-tain and episodic. The prevalent criterion was the same as in the case of early industrialisation, discussed above: the protection of natural resources and ecosystem integrity should not impose excessive limits on the freedom of locating and carrying out industrial production. This concern slowed down public action, allowing industry to ex-pand without the restraints of legislation protecting the reproduc-ibility of environmental resources. Even when, against much resistance, legislation to this effect was actually approved, it still left ample freedom to industries. A case in point is Act 615 of 13 July 1966 on environmental pollution. Instead of requiring industries to keep their emissions of a series of pollutants below fixed thresholds, it merely prescribed the adoption of measures to limit these emissions, leaving ample discretion in the bringing of plants up to standards, and only punishing offenders with small fines.

In spite of this, between the late 1960s and the beginning of the 1970s awareness grew, albeit timidly, of the impact that the expansion of urban-industrial systems was having on ecosystemic equilibria. The Fanfani government appointed a Senate committee for ecological issues. This was a first step towards the institution of the Ministry of Cultural and Environmental Heritage in 1974. Project 80 – that is, the preliminary report for the 1971–1975 national economic plan drawn up by the Ministry for the Budget and Economic Planning – set the guidelines for forms of land planning that would control urban growth and, at the same time, protect natural resources. One turning point was a long and detailed report on Italy's environmental situation drawn up by Tecneco, a company of the ENI group, and presented at a conference organised by the government in Urbino in 1973. This report marked the rise of a new awareness in public institutions and major private ones of the severe processes of deterioration and neglect threatening the country's natural resources and environment. It was an acknowledgement that economic growth was bringing pollution with it, whose consequences were especially serious in high population density areas in northern Italy and along the coasts of Lazio, Campania and Puglia. This report remained unmatched in scope and detail until the publication of the 'Relazione sullo stato dell'ambiente' (Report on the State of the Environment) by the then minister Giorgio Ruffolo in 1989.

Water pollution also became a matter of discussion and debate among technicians, entrepreneurs, local administrators and the national public powers. The Senate devoted a series of studies to the theme of the protection of water resources from pollution, published under the title *Problemi dell'ecologia* in 1971 and *I problemi delle acque in Italia. Relazioni e documenti* in 1972. In these studies, the issue of the scarcity of water resources was dealt with from the angle of economic planning and land management, with a special focus on the question of the costs of the restoration of contaminated natural resources. As we have seen above, Act 319 of 10 May 1976 (Legge Merli) finally launched a true water protection policy. This act imposed a uniform authorisation requirement on all discharges, independently of their use or purpose. Only in the 1980s, however, did the state of water resources improve significantly, thanks to the adoption of regional restoration plans and the activation of public depurators.

We should not forget that the 1970s were the years when the notion, however still vague, of preservation of natural resources for future generations began to take hold, starting from the Stockholm declaration on the environment adopted by the ONU conference. At the beginning of the same decade, EC legislation set underway a process of integration of the several spheres constituting the environment in order to develop protection policies. This process was drawn-out, finding its first full expression only as late as 1987, in the European Unico Act, which indicated the safeguarding of the environment and human health and the rational use of resources as objectives of the Community. An important turning point had been the EC's reception, in the mid-1970s, of the 'polluter pays principle', meant to dissuade polluters and stimulate research on tools and technologies with less environmental impact.

A search for an integrated concept of the environment also began in the Italian juridical debate. It was a long and complex process that cannot be gone into here. The difficulties arose from the multiplicity and diversity of the interests at stake and the conflicting rights claimed by different environmental stakeholders in the course of history. Interestingly, from the 1970s onward, the Italian constitutional court and Italian jurists tried to address these difficulties, gradually moving away from the traditional aesthetic characterisation of landscape towards a notion linked to that of environment, insofar as the landscape is part of nature and an integral component of the character of a given area. Since 2001, following the reform of Title 5 and the modification of Article 117, the Italian Constitution considers the environment jointly with cultural heritage, thereby overcoming the separation between the cultural-aesthetic and the ecological approach. This modification, however, has never ceased raising doubts, insofar as it transfers a large part of the state's competences in environmental protection and promotion to Regional, Provincial and town administrations.

From the 1970s onward, and especially in the 1980s, a different phase of industrial development began, characterised by a trend towards smaller and more scattered companies, especially in the Northeast. Industrial production grew further, although there were phases of slowdown. While energy consumption diminished, especially in the chemical and petrochemical industries, there was an overall growth in the flows of matter and resources absorbed by production. As regards pollution, the data are complex and vary from sector to sector. For example, nitrogen oxides and fine dust emissions decreased in sectors such as metallurgy and energy production, but increased in others, such as the food and clothing industries. The largest percentage of air pollution was caused by the energy industries. In general, there has been a growth of emissions of so-called 'greenhouse gases' – carbon dioxide, nitrogen oxides, methane and fluorides – as well as a growth of special waste, as we shall see below.

Act 3491 of 8 July 1986 introduced the concept of 'area at high environmental crisis risk', which marked a transition towards a systemic conception of pollution issues. Ministerial Act no. 471 of 20 October 1999 of the Ministry of Environment singled out 57 polluted sites requiring state intervention. These were mostly abandoned industrial areas, or still active ones requiring action to reduce their polluting impact. Some of these sites were more or less recently industrialised, notably Cengio, Porto Marghera, Caffaro in Brescia, Terni Papigno, Gela, Priolo, Taranto, Manfredonia, Porto Torres, Sulcis, the former Sitoco in Orbetello, Fibronit in Bari and the industrial area of Milazzo. Others had been chosen for the burial of special and dangerous waste. An epidemiological study by the Istituto Superiore di Sanità (Higher Institute for Public Health), known as 'Sentieri Project', analysed cancer statistics at 44 of these sites between 1996 and 2002, revealing higher than average mortality due to several types of water and soil pollutants. The emissions of refineries and petrochemical industries at Gela and Porto Torres, and metal works at Taranto and in the Sulcis-Inglesiente-Guspinese region, were responsible for major increases in deaths by lung cancer and non-cancerous respiratory diseases. At Massa Carrara, Piombino and Orbetello, heavy metals and halogenated components caused kidney problems. In Brescia, an increase in non-Hodgkin lymphomas was traced to diffuse contamination by polychlorinated biphenyls employed in the manufacturing of lubricants or as additives in paint, pesticides and other substances. At Balangero, Emarese, Casale Monferrato, Broni, Bari-Fibronit and Biancavilla, the presence of asbestos caused an increase in malignant tumors of the pleura. The Sentieri Project's studies showed that more than fifty per cent of the population of sites of national concern belonged to the poorer social classes. There was a growing awareness of the deep relationship between the social question and the environmental question. An economic growth that overlooks repercussions on ecosystemic balances, besides threatening to aggravate the costs bequeathed to future generations, accentuates social inequality by unequally distributing environmental risk.

2. Water management

After World War I, a group of technicians including Eliseo Iandolo, Angelo Omodeo, Carlo Petrocchi, Meuccio Ruini, Arrigo Serpieri and others – many of whom held institutional offices – developed a sweeping critique of the way water management works had hitherto been carried out. Their criticism mainly targeted the works carried out in the decades following national unification, especially in the South and Centre. As mentioned above, in the post-unification decades water management operations had been primarily carried out in the North, in the form of land reclamation in Emilia and Veneto and irrigation in Piemonte and Lombardy. In the South, although the investments in water management were double those allocated for northern and central Italy together, only isolated works had been undertaken; additionally, these works had failed to solve the problems they were meant to address, because several structural factors had been neglected in their planning, such as the region's hot and arid

Figure 3.1

The Ninfa garden before the start of the reclamation of the Agro Pontino, central Italy.
Taken sometime in the 1920s. Public domain photo from Wikipedia.

climate, the prevalence of mountain and hill settlements, far from work places, hydrogeological deterioration of mountain slopes, and the marshy conditions of the usually uninhabited plains. Malaria had been a constant of Italian environmental history. Caused by the bite of a mosquito of the genus *Anopheles*, which reproduces preferably in stagnant waters, it had strongly influenced the life and health of rural populations. At the dawn of unification, the areas most seriously affected by malaria were the Maremma, Lazio, the continental South and the islands. As late as the 1930s, coastal areas in the South were still malaria-ridden.

The Fascist period saw the rise of a new, integrated conception of land reclamation as simultaneously managing the hydraulics of plains and mountain slopes, improving hygiene and transforming the farming landscape. Significantly, the group of technicians who developed this new approach had collaborated with minister Nitti, who, as we have seen, as early as the beginning of the century had pursued a strategy built on land reform and the use of water and forest resources to produce electric energy and transform agriculture. The 'integrated water management' approach was implemented through a series of acts of law, such as Consolidated Act 3256 of 30 December 1923 on water management, Legislative Decree 753 of 18 May 1924 on the transformation of land of public concern and another act issued on 13 February 1933. These acts provided for the institution of consortia to carry out reclamation works, thereby expressing a will to make public and private interests go hand in hand. The largest operation was undertaken in the Agro Pontino countryside, south of Rome, where a huge land transformation was carried out from 1931 onward. The work was financed by the Opera Nazionale Combattenti (National Fund for Soldiers). Three thousand families who had come down from northern Italy were granted plots of the reclaimed land. By the end of the 1930s, 65,000 hectares had been transformed and parcelled out and five towns had been founded, including Littoria, present-day Latina, which became the Province capital.

This land transformation, however, was only fully accomplished in the decades after World War II. A package of extraordinary state measures helped to overcome the structural issues that had so far hindered the development of the Italian South. These measures were part of an overall project of transformation of the economy and society. Its objective was the management of water and forests, as well as land reclamation work, under a public programme made possible by the institution of a 'Fund for the South' (Cassa per il Mezzogiorno) and by an agrarian reform. Act 646 of 10 August 1950, entitled 'Istituzione della Cassa per opere straordinarie di pubblico interesse nell'Italia meridionale' (institution of a fund for special works of public interest in southern Italy), set the goal, as one reads in its first article, of implementing a programme of works inherent to 'the management of mountain basins and the watercourses they feed, land reclamation, irrigation, agrarian transformation, partly in connection with land reform, ordinary non-state roads, aqueducts and sewers, plants for the processing of agricultural produce, and tourist infrastructures'.

This programme was promoted by Prime Minister Alcide De

Figure 3.2

Reclamation of Parmigiana Moglia: collectors for the dewatering of the hoisting machine.
Public domain photo from Wikipedia.

Gasperi and strongly supported by a group of authoritative *meridionalisti* ('southernists') of several leanings and exponents of major public institutions, including Luigi Einaudi, Donato Menichella, Rodolfo Morandi, Nino Novacco, Pasquale Saraceno and others. In 1946, some of these men had founded the Associazione per lo Sviluppo dell'Industria nel Mezzogiorno (Svimez) (Association for the Development of Industry in the South), which became a major forum for a debate on policies and institutional resources to be deployed to improve the conditions of the South and bridge its gap with the North. The Cassa per il Mezzogiorno was created thanks

to a close cooperation that the governor of the Banca d'Italia, Menichella, managed to establish with the International Bank of Reconstruction and Development. This United States bank was willing to
give out loans with longer expiration dates than those granted under
the Marshall Plan. All these loans were to be transferred to a single
institution patterned after the one that, in the New Deal years, had
managed the development of the Tennessee river valley, which had
been hit especially hard by the crisis. This institution became a pillar of *meridionalista* culture. Later on, the Cassa per il Mezzogiorno
began action more specifically targeted at promoting the industrialisation of the South, notably with Act 634 of 29 July 1957, entitled 'Provvedimenti per il Mezzogiorno' (Measures for the South).
The act granted subsidies, loans and fiscal benefits for the purchase
of machinery and equipment, and provided for the creation of the
basic infrastructure required for the establishment of new productive activities. It also envisaged the institution of consortia of towns,
Provinces and chambers of commerce to set industrial development
under way. As Leandra D'Antone has stressed, between the late
nineteenth century and the 1950s and 1960s, prominent technicians,
scientists, bankers and managers played a crucial role in Italy's economic growth by promoting and inspiring the state's economic policies and implementing them through major institutions such as the
IRI and the Banca d'Italia.

The agrarian reform, instead, was largely carried out as a reaction to land occupation and increasingly harsh resistance from the
peasant movements after the end of the war. It was the final act in
a long-lasting history of conflicts over land. In the South, where the
after-effects of feudalism and the latifundium were stronger than in
the rest of the country, the land ownership question was especially
contentious. The reform consisted of parcelling land out among the
peasantry after it had been reclaimed and resettled. It was enacted
and regulated by Act 230 of 12 May 1950 (the so-called 'Legge Sila',
concerning vast areas of Calabria) the 'Legge stralcio', no. 841 of 21
October of the same year; and a Sicilian regional act issued on 27
December. This legislation was supplemented by acts granting tax and
credit benefits to facilitate land purchase, issued between 1948 and

1956. Thanks to these measures, between 1948 and 1960 the owner-ship of more than a million hectares of land, largely in the South and in the plains of central Italy, was transferred to peasant families.

As regards its effects on the environment, the reform changed the geography of smallholder property, which was now prevalently located in the hills and the plains. In 1961, smallholders held 52 per cent of hill farmland in the North, 24 per cent in the Centre and 42 per cent in the South, as well as 31 per cent of plain farmland in the North, 33 per cent in the Centre and 42 per cent in the South. While this process brought with it an improvement of living conditions and a democratisation of land property, it frequently had destructive ef-fects. The abandonment of the most elevated areas slowed and cut down erosion-preventing practices such as rainwater drainage, ter-racing and riverbed maintenance. This trend, along with unregulated building, increased hydrogeological instability.

The main results of these policy actions were land reclamation, improved sanitary conditions and the building of civil infrastructure. Partly thanks to new water adduction and irrigation technologies, hun-dreds of thousands of hectares of marshy land were reclaimed. This, along with the use of DDT, favoured the disappearance of malaria. The most important results were achieved in Campania, Puglia and Sicily. In Basilicata and Calabria, this renovation was harder to bring about due to the prevalently mountainous terrain of these regions.

The large plains of the South, the Tavoliere in Puglia, the Metaponto area, the plains of Sibari and Lamezia Terme, the Cro-tone area and the plain of Catania were definitively freed from ma-larial marshes and extensive cultivation, and given over to modern agriculture. In little more than twenty years, the surface of irrigated land in the South grew threefold, from 200,000 to 670,000 hectares. In the South, land reclamation and the peopling of the coastal plains led to a true 'reconquest of the sea', as Paolo Frascani called it. Its symbol is the *marine*, that is, the new seaside neighbourhoods of in-ner coastal towns, which started forming as early as the end of the nineteenth century. In Calabria, for example, by the 1950s there were more than fifty new *marine*.

Land reclaiming and irrigation brought with them a rapid and

substantial increase in agricultural production, especially in some regions. In Abruzzo, between 1951 and 1975 grain production grew
from 1.1 to 2.3 tons per hectare; in Sicily from 1 to 1.9; in Puglia
from 1.09 to 2.7. This led to a significant change in the geography of
agricultural production in the Italian South. Grain was superseded by
intensive vegetable and fruit growing, especially in certain areas, such
as the Agro Nocerino-Sarnese, the former Tavoliere di Puglia and the
plain of Catania.

An industrialisation of the South got under way. It involved
both private and public companies, most notably the chemical industries of Augusta-Priolo-Melilli in the Province of Siracusa and in
Brindisi, Gela and Manfredonia; Olivetti at Pozzuoli; the iron works
of Bagnoli and Taranto; and the Alfasud car industry at Pomigli-
ano d'Arco. Mechanical, electronics, telecommunications, textile and
clothing companies – some funded with foreign capital – were established in Campania, especially in the Caserta and Naples area.

No univocal judgment can be given on the extraordinary policy measures for the South. In the initial stage of their application,
their effect was undoubtedly positive. They provided rural areas in
the South with a vast network of civil infrastructure – aqueducts,
electricity, sewers – and freed them from the plague of marshes and
malaria. The social transformation brought about by the agrarian reform was also positive. The reform dismantled the latifundium and
the feudal-type relations it was founded on, although the small extension of the new holdings sometimes hindered them from evolving
into businesses capable of facing the challenges of the market. As to
the subsequent infrastructure-building stage and the ways in which
the industrialisation of the South was promoted, the judgment has
been more controversial. The sociologist Carlo Trigilia defined the
effect of the extraordinary policy actions as 'development without
autonomy' (*sviluppo senza autonomia*), as these actions were incapable
of activating an endogenous growth.

As Ada Becchi has stressed, from the 1960s onward the infrastructure policy of the Cassa per il Mezzogiorno eventually became
a means to attract financial resources in order to build consensus and
fuel a patronage system. Roads, for example, were planned and built

regardless of the existence of an actual demand for transportation and access, and ignoring traffic and pollution issues and the impact on the environment and landscape.

Indeed, the political culture of *straordinarietà* ('extraordinariness') had serious negative environmental repercussions. It did not merely encourage local administrations to delegate their responsibilities to the state; it eventually widened the gap between a decision-making centre and a myriad of areas lacking solid local government and standing at the mercy of more or less licit private interests. This trend became especially strong from the 1980s onward. During this period, collusion between administrators, politicians and criminal organisation grew and became endemic, with destructive effects on the environmental balance of many areas.

3. The perception of the environmental crisis

In the second postwar period, while historic associations such as the Touring Club and the Club Alpino expanded their membership, a variegated array of environmental associations arose, bearing witness to a growing awareness of environmental issues. In 1955, a group of intellectuals including Umberto Zanotti Bianco, Giorgio Bassani, Antonio Cederna, Elena Croce and Desideria Pasolini dall'Onda founded Italia Nostra. This association, whose members were prevalently architects, urbanists and intellectuals, was especially active in the defence of artistic and cultural heritage and in the protection of historic town centres and the rural landscape. Over the years, it broadened its interest to include more strictly naturalist concerns, constituting the Gruppo Verde (Green Group) and proposing a framework law on national parks in the wake of the First World Conference on National Parks held in Seattle in 1962.

The struggles of Italia Nostra intersected with those of authoritative exponents of journalism who, from the pages of newspapers such as *La Stampa*, *l'Unità*, *L'Europeo*, and *L'Espresso*, helped to boost public awareness of environmental issues. The architect Antonio Cederna was one of the most active and committed exponents of Italia Nostra. From the pages of Mario Pannunzio's *Mondo* and later on from those of the *Corriere della Sera* and *Repubblica*, he fought important battles against speculative building, the tearing down of historic centres, the destruction of valuable landscapes, the expansion of rundown suburbs, land rent and the inadequacy of urban planning. Italy was growing richer at the expense of its environment. Its growth was fed by a land market guided solely by the speculative logic of subjects whom Cederna called 'the thugs of concrete', including the Società Generale Immobiliare di Roma, one of the main targets of his accusations.

On the more strictly naturalistic side, 1948 saw the foundation of the Italian Movement for the Protection of Nature, followed by the National League Against the Destruction of Birds in 1965 (later renamed Italian League for the Protection of Birds). Most notably, the year 1966 saw the inauguration of the Italian section of the World Wildlife Fund, whose main animators were Arturo Osio and Fulco Pratesi, members of Italia Nostra's 'Green Group'. After the example of the British National Trust, this association aimed at constituting natural oases after having purchased or rented land. Although they did not manage to gather as many members as their foreign counterparts, in the course of the 1970s both Italia Nostra and the WWF significantly increased their membership; that of the former doubled (from 10,000 to 20,000) and that of the second trebled, eventually growing to more than 30,000. Other important organisations emerged in those years. One was the Federazione Nazionale Pro Natura, which was strong in all the Italian Regions and in 1970 incorporated the Movimento Protezione della Natura. Another was the FAI (Italian Environmental Fund), founded in 1975 to protect Italy's artistic and landscape heritage.

In the 1970s, the theme of the depletion of natural resources due to the extraordinary growth of consumption took centre stage in the public debate. This was largely a result of the publication of the study *The Limits of Growth* in 1972. This report was drawn up by a group of MIT scholars upon invitation from the Club di Roma (Rome Club) – an informal forum of public officials, scientists and managers. Aurelio Peccei edited and promoted the book. In the same year 1972, the UN conference on human environment was held in Stockholm, attended by representatives of 113 countries.

Figure 3.3

Flooding in Florence. Looking Towards Piazza Santa Trinita, Florence, November 4, 1966. Photo from the Balthazar Korab archive in the Library of Congress in Washington D.C.

On that occasion, the Declaration of the United Nations Conference on Human Environment was drawn up, which affirmed the principle that the defence of the environment is a priority concerning the future wellbeing of all the people of the Earth. The report was part of a more general reflection on economic backwardness, the North/South gap, overpopulation and world hunger, which had gone hand in hand with decolonisation in the second postwar period. In 1962, the American biologist Rachel Carson published *Silent Spring*, which later became a cult text among environmentalists. The book denounced the destructive effects of chemical products and DDT on crops and, more generally, on ecosystemic balances. The theme of development thus became increasingly connected to that of the reproducibility and depletion of natural resources.

The Rome Club report explained that the growth of agricultural and industrial production consequent on the increase of world population, besides accentuating pollution and soil depletion, would eventually result in reduced availability of natural resources and this would lead to an increase in disease and death, and in conflicts over scarce resources. Hence the need to limit economic growth. The report, which was translated into fifteen languages, had an extraordinary impact and sparked discussions and arguments that were to last until the present day. In Italy, it did not garner much praise among economists or in the industrial world. It was also harshly criticised by exponents of the Left, who apparently saw it as an expression of the will of international capitalism to overcome its contradictions through a sort of technological authoritarianism.

In the meantime, increasingly often the media reported cases of environmental pollution and harm to human health produced by waste dumping and industrial emissions. These included the effects of chemical productions in the Bormida river, at Porto Marghera, in the Mincio near Mantova and at Manfredonia, as well as the marble powder pollution in the quarries of Versilia, the contamination of the Lambro and Olona rivers and the damages caused by the ironworks at Taranto and Bagnoli. Increasing public awareness of the serious environmental impacts of industrial production had favoured the spread of new environmentally-conscious professional groups such as Medicina Democratica and Geologia Democratica, and the foundation of journals such as *Ecologia* (later renamed *La Nuova Ecologia*) and *Sapere*. In the same years, some major works were published, notably *The Environmental Revolution* by Max Nicholson – one of the founders of the WWF – in 1971, and *Ecologia e lotte sociali, ambiente, popolazione, inquinamento*, by Virginio Bettini and Barry Commoner, in 1976.

Awareness of environmental issues also grew among magistrates. The actions of 'assault praetors' (*pretori d'assalto*), as some prosecutors of the time were called, marked an important phase in the history of environmental and health defence in Italy. While they could not remove the deeper causes of environmental destruction, these magistrates did contribute to improve resource management, sustainability and legality of resource use, and thus tried to make up for the slowness and inadequacy of political action.

In the Left – as the above-mentioned reactions to the Club di Roma's report bear out - the ecological question was perceived in conflicting ways. At a conference held at the Istituto Gramsci in 1971, Giovanni Berlinguer stressed that the international workers' movement was lagging behind on environmental issues and argued that ecology was not merely one more problem to deal with, but a new dimension of politics. Conversely, in 1972 the Marxist Dario Paccino – the author of the book *L'inganno ecologico* (1972) – expressed the diffidence of much of the Italian Left towards ecology, which he saw as a means for the bosses to get rich through pollution-reducing industrial productions, without challenging the system as a whole.

The Left never regarded the environmental question independently of concerns over unemployment. Nevertheless, prominent members of the Italian Communist Party embraced environmentalism. An especially noteworthy one was Laura Conti, an authoritative exponent of scientific environmentalism. Besides developing a critique of science and development's dissipation of natural wealth without actually meeting people's real needs, as a physician Conti struggled for the enforcement of measures to curb pollution in various sectors, from detergents to insecticides, from vehicle traffic to plastic. Another scientist who was active on the ecological front – as a member of Parliament elected as an independent campaigner in the Communist Party – was Giorgio Nebbia, a disseminator of scientific environmentalism who formulated an ecological critique of capitalism.

Although the environmental question remained a minor concern of the Communist Party during the 1970s, the political proposal put forward by its secretary, Enrico Berlinguer, in 1977 actually had a lot to do with environmentalism, insofar as it promoted an alternative to a capitalist system based on unlimited increase of consumption and on the squandering of natural resources. The limits of this system had been exposed by the major petrol crisis of 1973. Berlinguer's proposal was based on an awareness that the entry of former colonial peoples and countries into the world arena undermined existing equilibria and highlighted capitalism's inability to make up for the deep economic and social inequalities between developed and underdeveloped countries, and within the same country. In the following decades, this analysis found an echo in a critique of development measured exclusively in economic terms and in the consequent idea of a form of human progress where what counts is health, education, environmental quality and freedom.

In the composite environmentalist universe that was emerging in the 1970s, the culture of 1968 played an important role. In this culture, the discussion of environmental issues intertwined with a critique of technical-scientific knowledge and its subordination to the dominant economic interests of capitalist society.

At the same time, concerns over the health of workers and industrial pollution were rising. The problems of the noxiousness of the working environment had so far remained marginal in trade union claims. At the time, environment and employment appeared as incompatible concepts, as did defence of nature and economic development. A revealing example of this is the case of Icmesa, a chemical plant which in 1976 emitted a toxic cloud that contaminated much of the territory of the town of Seveso, in Brianza. The local population and the plant workers exhibited a certain reluctance to denounce the responsibilities of the company and to mobilise in defence of their health; rather, they chose to downplay the seriousness of the incident for fear of jeopardising their jobs.

During the same years, the energy crisis that followed the petrol shock of 1973 persuaded Western governments to set their stakes on nuclear energy. In Italy, as elsewhere, this decision provoked the rise of an antinuclear movement. The plan to build a nuclear power plant at Montalto di Castro, north of Rome, induced this movement to take a more decisive stance and thus gain centre stage in the public debate. The mobilisation against the power plant, which went on until the second half of the 1970s, drew a variegated array of participants, ranging from the WWF, groups of the 'New Left', the 'Movement of '77', the Radical Party, the Amici della Terra (Friends of the Earth) association – founded in 1977 and very close to the Radical Party – Democrazia Proletaria, the Federazione Unitaria Lavoratori Metalmeccanici and Unione Italiana dei Lavoratori trade unions, the Associazione Ricreativa e Culturale Italiana and some exponents of the Socialist party. The Italian Communist Party, however, retained its pro-nuclear stance. This position was partly revised in the following years, after Fabio Mussi and Antonio Bassolino put forward a motion against nuclear energy at the 1986 congress; their motion, however, gathered more support in regional congresses than among the party's national management. In 1978, a committee for the assessment of energy policies was constituted. This committee played a leading role in a movement of antinuclear scientists and technicians – including Virginio Bettini, Marcello Cini, Gianni Mattioli, Giorgio Nebbia and Massimo Scalia – that was to become a beacon of Italian environmentalism. In the early 1980s, this first phase of mobilisation lost its momentum, but picked it up again in 1986–87, when the Chernobyl plant accident and the consequent victory of the referendum to abolish nuclear plants spelled the end of the nuclear project in Italy.

4. Policies for cities

Over the last two centuries, the world has seen an impetuous development of cities. Urban population has grown from ten per cent of the total in 1800 to almost half by the end of the twentieth century. Today, eighty per cent of the world gross national product is generated in cities. Metropolises have sprung up, large cities with immense suburbs, whose surroundings are insufficient to meet their food and energy requirements, and whose connections hence extend outside of their limits, even to remote areas. Such an urban regime involves an extraordinary consumption of energy and natural resources, and a dramatic change in land use and forms of pollution.

In the present situation, where cities have become one of the main factors in pressure on environmental equilibria, urban land management policies have gained crucial importance. Only public management can limit the damages that senseless land consumption, a lack of services and infrastructures and unregulated contiguity between industrial and residential areas can wreak on the environment, historical and natural heritage and citizens' wellbeing.

Especially significant urban management experiences came to maturation in England, France and Germany as early as the nineteenth century to cope with the dramatic changes brought on by the unbridled expansion of capitalism. In Italy, general norms defining principles and procedures valid for the whole of the national territory were not issued before the urban planning law of 1942, which obliged all town governments to approve their own town plan. Before this date, individual expansion plans, usually implemented as state legislative acts, were launched for large and medium-sized towns such as Rome, Milan, Naples, Florence, Turin and Genoa.

After World War II, the extension of the urban-industrial system and the rise of an energivorous society produced an unprecedented impact on Italy's landscape. Until the 1970s, town plans were the tool used to programme urban expansion areas and infrastructures, especially roads. The main issue in the public debate on land use was the boundary between the rights of private citizens and those of the collectivity. In 1962, the Christian Democrat minister of public works Fiorentino Sullo sponsored an urban reform bill that intended to dissociate the right to build from property rights and delegate the right to exercise it to public authorities, which were to regulate building through their town plans. The bill envisaged general and preventive expropriation of buildable areas and compensation for expropriation, based on the agricultural value of the land. The purpose was to curb speculative building. After seizing the building areas, town governments would have undertaken primary urbanisation works and subsequently given over the areas destined for residential construction to private subjects, while retaining their property.

The reform bill caused lacerations and conflicts even within the government parties, insofar as it appeared as a limitation to owners' freedom to use their land that went against their interests. According to Vezio De Lucia, fear of an urban reform going in this direction was one of the motives behind the attempted right-wing coup of 1964. The bill, at any rate, was dropped, leaving a phase of intense urbanisation unregulated. The reform was later enacted, albeit only in part, with the Bucalossi Act of 1977. There is no doubt, however, that the failure to approve the Sullo reform gave free rein to speculation in urban growth – with buildable areas being bought at very low prices and resold at artificially high ones – and destruction of natural equilibria and the landscape. The cities of the economic miracle had slowly expanded in all directions into the countryside, often neglecting to reserve areas for public parks and services. Construction has been one of the most lucrative sectors in Italy, thanks to the increase of urban population and the consequent growing demand for buildable land. The construction sector has been at once one of the drivers of the Italian economic growth and one of the main factors in Italy's environmental imbalance.

The failed urban reform is a reflection of the Italian state's historical weakness vis-à-vis private interests, which explains the recurrent difficulty of imposing a public regulation on ecosystemic and environmental balances.

In the intense and lively debate of those years, when the theme of public land control went hand in hand with that of housing and urban livability, the so-called 'reformist town planning' arose, largely spearheaded by the National Institute of Urban Studies. This approach differed from the traditional one because it did not limit itself to functional organisation of the urban space, but proposed to reduce the negative consequences of urban development by addressing structural issues through control of the real estate market and the mechanisms presiding over the formation of rent.

Exponents of this 'reformist' trend included Guido Alborghetti, Felicia Bottino, Pier Luigi Cervellati, Giuseppe Campos Venuti, Edoardo Detti, Vezio De Lucia, Edoardo Salzano, Alberto Todros and many others. They were linked both with public administrations and with left-wing parties, and were behind many institutional decisions made both at the national and at the local level between the second half of the 1970s and the 1980s.

The centrality gained by urban studies in the political and institutional debate of the 1970s belongs within the broader framework of the reformist culture typical of social democrat tradition. This culture informed the economic planning policies adopted in that decade, such as the preliminary report for the 1971–75 national economic programme or the Progetto 80, which aimed to make up for economic and infrastructural gaps between different areas in the country and promoted a town-planning model envisaging 'city systems' to keep in check the most adverse consequences of spontaneous urban expansion.

Reformist town planners also endorsed a new conception of urban development founded on refurbishment and restoration rather than expansion. Increasing attention was placed on the issue of historic town centres, an issue that rightfully belongs within the purview of environmental history, especially in Italy, where environmentalist movements have always been concerned with both natural and cultural heritage, with historical as well as natural identity.

From the 1960s onward, views about the protection of historical town centres from demolition and gutting changed deeply. Whereas up to then this protection had focused on buildings of special artistic value designated as 'monuments', henceforth it was extended to the whole historical fabric of urban centres. The Gubbio Charter, issued at the conclusion of a meeting promoted by the Associazione Nazionale dei Centri Storico-Artistici in 1960, affirmed the principle that restoration and renovation plans should concern not only the buildings per se, but also the people who lived and worked in them; in other words, after renovation was completed the buildings should be given back to their original inhabitants.

The 1960s and 1970s thus witnessed a number of policy actions to restore old neighbourhoods and their surroundings, including the planning of the historical centres of Siena and Matera, the protection of the hills of Florence, the Assisi town plan, the Venice area plan, the Fora plan in Rome and the plan for the suburbs of Naples, approved at the end of the 1970s and enacted in the framework of the reconstruction that followed the 1980 earthquake. The plan for the renovation of the historical centre of Bologna, drawn up by Pier Luigi Cervellati and adopted in 1977, remains the principal model for this culture of renovation. This plan was inspired by the notion that urban conservation and restoration should also include natural features, drawing on a conception of cities as dualities of nature and artefact, habitat and society.

Environment in Republican Italy

The issue of historical centres is also connected to the quality of urban life and to housing policies. The new town-planning approach has shown that the restoration of existing buildings can curb land consumption, resource waste and unregulated building.

In those years, housing was a hot issue. The ways in which the country had developed – industrial polarisation and the consequent concentration of migratory flows, an urban expansion characterised by a high demand for middle and mid-lower class housing versus a supply of mostly luxury homes – had generated an explosive social situation. Several cities – including Milan, Turin, Rome and Naples – saw the rise of strong housing rights movements, in which the trade unions and students' political organisations often found themselves side by side. Starting from 1968, the trade unions displayed a special interest in the struggle against building and land speculation, in state action to build new housing and in more efficient urban policies.

Important legislative acts addressing these issues were approved during these years. Act 865 of 1971 regulated the planning of policy actions and expropriations for the common good, carrying forth and reinforcing a reform that had already been set under way in the previous decade with Act 167 of 1962. Act Bucalossi of 28 January 1977, mentioned above, regulated the relationship between land ownership and the right to build. The equitable rent (*equo canone*) Act of July 1978 protected home rents from the vagaries of the free market. Furthermore, a ten-year housing plan issued in August 1978 provided for the use of public funding to restore existing buildings.

CHAPTER FOUR.

ENVIRONMENT AND ENVIRONMENTALISM IN CONTEMPORARY ITALY

1. Land protection and land consumption

As we have seen above, the depopulation of mountain and hill areas accelerated Italy's hydrogeological deterioration and was thus one of the major causes of the disruption of the country's environmental equilibria. The other main cause was urbanisation. Ever since the 1950s, land consumption has been growing at a much faster pace than population. Italian towns have expanded into the countryside without the least concern – with a few exceptions – for maintaining green belts, which other European countries established as early as the aftermath of World War II, following the principle that town planning should define the relationship between the urban and the rural space, and set limits to city growth for the sake of citizens' health and wellbeing.

The growth of Italian towns – large, medium and small – has been a random and unregulated process, which eventually not only devoured valuable farmland and landscapes, but also encroached into areas at high hydrogeological risk. These areas, as we have seen, are intrinsically fragile due to the character of their soil and hydrographic systems. Waterproofing of the ground, the deviation of watercourses and the facing of their beds with concrete, the building of housing on steep slopes and extraction of materials from torrent beds have accentuated hydrogeological risk. Additionally, the advent of the energivorous society has placed land under new forms of pressure; for example, the development of the transportation network involved the digging of shafts and tunnels, with detrimental effects on groundwater reservoirs. In the second postwar period, landslides and floods grew dramatically in Italy. According to data of Legambiente, today about 82 per cent of Italian towns are at hydrogeological risk. A paradigmatic instance of the lack of a land conservation culture

Figure 4.1

The Vajont Valley after the Vajont dam disaster.
Public domain photo from Wikipedia.

in the Italian modernisation is the Vajont disaster. In October 1963, a huge landslide fell into an artificial hydroelectric basin built in a geologically unsuitable area, generating a wave that spilled over the dam and ran down to the valley bottom, killing almost 2,000 people. There were also disastrous inundations of the Reno and Po, floods in Sicily, Sardinia and Calabria, and a major flood in 1966, which devastated extensive parts of central Italy, including Florence. The list goes on with the Agrigento landslide of 1966 and the Genoa flood of 1970, both aggravated by scarcely planned urban expansion into particularly vulnerable zones. Then there were the Sarno mudslides of 1998, and floods in Calabria and Piemonte in 2000, in Genoa, again, in 2011, in Massa and Carrara in 2012, in Sardinia in 2013 and in Genoa for the third time in 2014.

A different kind of threat was generated by the uncontrolled urbanisation, mostly unlawful, of a strip of land at high eruption risk around Mount Vesuvius, the so-called Red Zone. An evacuation plan has been drawn up for this area.

The 1980s marked a turning point. During this decade, much legislation was issued to create tools for safeguarding environmental equilibria and protecting the landscape. The Galasso Act of 8 August 1985, no. 431, entitled 'Disposizioni urgenti per la tutela delle zone di particolare interesse ambientale' (Urgent measures for the protection of areas of particular environmental interest), recognised the environment as an asset of public interest. The earlier legislation, which was based on the Act of 1939, ruled that an area with specific landscape and environmental features could be placed under restrictions, for example by forbidding construction on it. The Galasso Act, instead, protects landscape elements as bearers of values per se and globally. It places a general restriction on a 300-metre strip along the seaside and the banks of lakes, rivers and waterfalls, around the highest parts of mountains, glaciers and volcanoes and around areas of archaeological or forestry interest. Furthermore, it can be enacted both in the town plans of Italian regions and in urban plans with a special concern for the environment.

As to hydrogeological instability, which had so far been addressed with ad hoc measures, Act 183 of 1989, entitled 'Norme per il riassetto organizzativo e funzionale per la difesa del suolo' (Norms for organisational and functional restructuring for the defence of the soil) introduced structural measures through hydrographic plans drawn up by basin authorities. This act affirmed the principle that, to effectively address both environmental and social issues, public planning must encompass the ecosystemic boundaries of an area rather than its administrative ones. It rules that basin plans should provide binding directives for individual urban town plans in order to harmonise the defence of the soil and urban organisation, especially as regards the mapping of hydrogeological risk.

Figure 4.2

Flooded olive grove, Sicily. Photo: Stefania Bonura.

The importance of Act 36 of 1994 on water resources, known as 'Legge Galli' (Galli Act), lies especially in its extension of the 'public water' category to superficial underground water. This act also envisages an integrated management cycle of water simultaneously meeting users' needs and recycling water, thereby reducing the destructive impact of the drawing of groundwater and the depletion of riverbeds. Act 10 of 1991 ruled that the general urban plans of towns with a population higher than 5,000 should include a section relative to the use of renewable energy sources. Law 225 of 1992 instituted the National Civil Protection Department, which coordinates rescue activities during natural calamities, as well as striving to foresee and forestall them. Finally, Decree 180 of 11 June 1998, issued in the

Figure 4.3

Landslide, Sarno, southern Italy, 1998. Source: public domain image from http://www.meteoweb.
eu/2012/04/alluvione-di-sarno-e-quindici-14-anni-dopo-in-memoria-dei-160-morti-sotto-il-fiume-
di-fango/131887/

wake of the May 1998 catastrophe in Sarno and its neighbouring towns, and converted into Act 267 in the same year, was a plan to ensure safety in the areas at highest risk. The assessment of this risk was left to the Ministry of the Environment. Today, thanks to the institution of 25 new national parks under Act 394 of 6 December 1991 on protected areas – only two had been established previously since the end of the war, that of Calabria in 1968 and that of the Aspromonte in 1989 – with the addition of a number of regional parks, about ten per cent of the national territory is under nature protection, which involves special protection of the hydrogeological balance, conservation of plant and animal species and promotion of artistic and archaeological heritage.

The legislative turning point of the 1980s was the result of an institutional process that had started decades earlier. The two green plans for agriculture of 1961 and 1966 envisaged the sowing of crops that would protect the soil from water erosion. The national economic programme of 1966–1970 recognised that the issue of soil defence should be part and parcel of economic policy guidelines. An interdisciplinary committee for hydraulic management and soil defence instituted by a state act of 1967, known as 'commissione De Marchi', went so far as affirming the principle that the protection of land from hydrogeological calamities is a vital interest of the state and that all actions undertaken for this purpose should encompass hydrographic basins as a whole.

The 1980s thus marked a major turning point in terms of awareness of the public character of land management issues and in terms of the rise of the notion of the environment as a collective asset; at the same time, however, there was strong resistance against the new legislation in society and in public institutions. One only needs to think of the reluctance of town administrations, especially in some areas of southern Italy, to adopt town plans (in 1978, only 193 out of 2,522 southern towns had one), of the spread of unregulated building, of the frequent criminal infiltrations into local administrations (especially in areas with strong traditional Mafia presence) of the many conflicts of responsibility between local administrations, of the gradual devolution of state responsibilities to the Regional governments – whose technical and scientific know how did not always prove up to the task – and of the inadequacy of control, partly due to the weakening of institutions such as the state civil engineering department (Genio Civile) and the public works departments. Furthermore, the newly established institutions for nature protection imposed restrictions, but rarely followed them up with efficient protective legislation and coordinated land management decisions. This encouraged frequent recourse to exemption clauses and widespread resentment against public planning.

As regards urbanism policies, the 1980s began with the dismantling of the legislation regulating the relationship between property rights and building rights. This was followed, in 1985, by a general conditional amnesty for building law infringements (*condono edilizio*) committed until October 1983. A trend to the liberalisation of land use thus resurfaced, which constitutes a characteristic trait of Italian history but gathered new strength during the years of the *pentapartito* (the five-party coalition government) and the two Berlusconi decades. This trend is significantly reflected in deregulation measures such as another conditional amnesty for building law infringement issued as Act 326 of 2003 – whose expiration term was repeatedly extended until the end of 2004 – and the housing plan of 2009, meant to revitalise the private building sector by allowing citizens to bypass town plan restrictions on home expansion. In many large towns, forms of 'contracted planning' prevailed, which left room for private initiative and delegitimised the role of public planning. (Naples was an exception in this regard; in 2004, after a drawn-out approval process, it issued a town plan entirely conceived by the city urban planning department.) In Milan and Rome, in particular, planning was not guided by a general vision, but was in fact controlled by power groups born of alliances between real estate businessmen, financial institutions and public administrations.

In the 1990s, the trend to housing dispersal grew, as did the pace of land consumption for urbanisation. Between 1990 and 2006, changes in land use affected a total of ca. 552,000 hectares, equal to the surface of the Liguria Region. Urban areas increased by 131,300 hectares and forests by 84,800. Land consumption for urbanisation grew by eighteen per cent in the mountains and 88 per cent in the plains. In hilly areas, the growth was forty per cent for forestland and 44 per cent for urbanisation. The rate of land consumption was also high along the coasts. According to a WWF study conducted between 1995 and 1997, 58 per cent of the Italian coast was densely built up, thirteen per cent was occupied by scattered buildings, and only 29 per cent was still free, although partly occupied by camping sites, greenhouses and fish farm buildings.

The highest increases in urbanisation were recorded in Piemonte, Lombardy, Veneto, Emilia-Romagna, Tuscany, Calabria and Sardinia. This growth took the form of urban sprawl, which is to say that it was not merely the result of a natural expansion of city suburbs, but was characterised by chaotic dispersion and high land consumption by new settlements, accentuating illegality and social marginalisation. The shrinking of productive and natural land was not the only environmental consequence of the expansion of urban sprawl. Another obvious effect was an increase of almost exclusively private vehicle transportation, due to the impossibility of providing adequate public transportation for entire new town neighbourhoods. In large metropolitan areas, urban sprawl has produced destabilisations and criticalities that have undermined the quality of life and of the environment. Towns like Milan, Rome and Naples have witnessed similar expansions of their hinterlands, which are administratively fragmented and increasingly polluted and degraded. The absence of a single guiding body has impeded the enactment of consistent policies, especially as regards services. The expansion of areas lying outwith public transport networks and far-removed from workplaces has resulted in large-scale private-vehicle commuting. This phenomenon has proved especially dramatic in Rome, where public transportation is especially inadequate. With 67 cars per 100 inhabitants, Rome is the second most motorised city in Italy, after Catania, and one of the most motorised in Europe. The institution of the 'città metropolitane' by Act 57 of 7 April 2014 – the so-called 'Delrio Act' – is meant to invert this trend by reorganising fourteen metropolitan areas in such a way as to meet the population's needs across town administrative boundaries.

The car industry was the driver of Italy's whole industrial system. Its growth and the spread of private vehicles cannot be explained exclusively in terms of economic categories such as income and consumption growth. A symbol of autonomy and prosperity, in Italy the car has been an important facet of a more general social transformation, as Federico Paolini has shown. Between 1961 and 1995, the number of cars increased steeply, especially in future metropolitan areas: six-fold in Rome, Milan and Turin, and as much as eleven-fold in Naples. Throughout the 1990s, the average Italian growth rate of car ownership was higher than the Western European average, 48.3 vs 39.3 per cent in 1990 and 56.3 vs 45.7 per cent in 2000. As a consequence, road traffic has become one of the principal causes of air pollution – as indeed it is worldwide. As a measure of the magnitude of the increment of carbon dioxide emissions – the prime cause of global warming and global climate change – in Italy between 1870 and 2000 emissions went up from 3,127 to 443,738 thousand tons. Although the highest peaks were in the 1960s and 1970s, carbon dioxide emissions have not ceased to grow in the following decades, albeit at lower rates.

2. Political environmentalism

As we have seen, in the 1980s a decisive phase opened in the history of environment, both in terms of rising public awareness of environmental issues and in terms of the effects of this awareness at the political and institutional level. The year 1986 was especially important, as it saw the creation of separate ministries for the environment and cultural heritage, and the foundation of the National Federation of Green Lists, the outcome of a process started in 1981, when the Arcipelago Verde (Green Archipelagos) – a national coordination of ecologist and nonviolent associations, committees, journals and radio stations – was established in Bologna. In a meeting organised in Trento in 1982, Alexander Langer, Regional councillor for the New Left, came to the fore as a leader of this movement. The Green Federation mainly gathered its vote among the electorates of the extreme-left parties, the Radical and Socialist Parties, the center lay parties and in part even the Republican Party. With the foundation of its Green Party, Italy followed in the footsteps of European countries with a deeper and longer-running history of environmentalist concern. The Ecology Party had been founded in the United Kingdom as early as 1975, the Verts in France in 1982 and the Grünen in Germany in 1980; this latter organisation constituted a model for ecologist movements all over Europe.

In the same year, 1986, the explosion of a reactor in the nuclear power plant of Chernobyl in Ukraine, on 26 April, caused the formation of a radioactive cloud that extended even to faraway areas, including Italy. The event produced a deep change in world public opinion, stimulating awareness of the existence of an ecosystemic crisis of planetary dimensions and of the transnational character of environmental risk.

In a referendum held on 8 and 9 November of the following year, eighty per cent of Italian voters (65 per cent of the eligible electorate) voted against nuclear power plants. This spelled the end of the use of nuclear energy in Italy. Over the years that followed, all of Italy's nuclear plants – located at Trino Vercellese, Caorso in the Province of Piacenza, Borgo Sabotino in the Province of Latina and Garigliano near Caserta – were shut down. 1987 was also the year when Italian environmentalism managed to send representatives to Parliament. In 1985, the Greens had run for local elections in eleven regions under the symbol of the 'Laughing Sun', winning a total of 140 Regional, Provincial and town councilors. In the 1987 election, the Green Federations obtained 2.51 per cent of the vote in the Chamber of Deputies and 1.96 per cent in the Senate, which earned them thirteen Deputies and two Senators. Other exponents of the Green universe were elected to Parliament in the lists of other parties, notably Laura Conti and Chicco Testa with the Communist Party, Giorgio Nebbia and Enzo Tiezzi with Sinistra Indipendente and Edo Ronchi and Gianni Tamino with Democrazia Proletaria.

The political ascent of the Italian Green Party, although less impressive than that of other European ecologist parties, was confirmed in the European election of 1989, in which the Greens ran as part of a coalition that chose the name of Verdi Arcobaleno (Rainbow Greens). This coalition, which was close to the Italian Communist Party, also included the Lega per l'Ambiente (League for the Environment) and the Lega Italiana Protezione Uccelli (Italian League for the Protection of Birds). It managed to elect five members to the European Parliament, having garnered 6.2 per cent of the Italian vote (ca. 2,000,000 votes). The years between 1987 and the four-year span during which Edo Ronchi was minister of the environment (1996–2000) were undoubtedly successful for Italian political environmentalism. During this period, part of the objectives the party had pursued during its movement phase were actually enacted by a series of legislative acts, viz., in 1989 the Act for the defence of the soil; in 1991 Acts 9 and 10 on the promotion of renewable energy sources and energy saving, as well as Framework Law 394 on protected natural areas; in 1992 Act 257 banishing asbestos and an act allowing hunting only when it did not threaten conservation of the fauna; and in 1994 the Galli Act on water management, as well as an act instituting a national agency for the protection of the environment(Agenzia nazionale per la protezione dell'ambiente), whereby environmental control was entrusted to the Regional governments through the Arpa Regional agencies. In 1995, the first bicameral committee for investigation on the waste cycle and illicit activities connected to it was instituted and placed under the presidency of Massimo Scalia. In 1997, the Ronchi Decree on the reorganisation of the waste disposal sector was approved.

In the meantime, the environmental question had attained transnational and planetary dimensions, and was being addressed by large international organisations. In June 1992, the UN Conference on Environment and Development was held in Rio de Janeiro, where it approved the convention on climate and biodiversity. In December 1997, the protocol against climate change was signed at Kyoto. It came into full force in 2005. In August 2002, a world summit on sustainable development was held in Johannesburg. Five years later, the 27 countries of the European Union signed a binding pact to reduce greenhouse gases by twenty per cent by 2020. In December 2009, a UNO summit was held on climate changes caused by rising temperatures as a result of deforestation and the increase of greenhouse gases in the atmosphere.

Political environmentalism spread from the West to the rest of the world. Green parties sprang up in some Asian, African and Latin American countries. In New Zealand, ecological parties had existed as early as the beginning of the 1970. There was thus a worldwide mobilisation, which favored global environmentalist initiatives. In 1999, the anti-global movement arose in the United States, during the World Trade Organisation summit. In 2001, the first World Social Forum was held in Porto Alegre.

Over the last few decades, environmentalist awareness and concerns over landscape and land protection have gained hold in civil society. Citizen initiatives have multiplied, largely as the result of the work done by the world of associations and some town administrations. Such initiatives include beach cleaning rallies, checks on seawater quality along the coast, forums and campaigns against pesticides, public petitions on issues such as food security and demonstrations against polluting emissions in cities and around factories. In its struggle against the destruction of Italian fauna, the WWF scored major results, such as the reintroduction of the red deer, the roe deer and the lynx, which had become almost extinct.

Environmentalist literature also flourished in this period, including some books that were to become beacons for environmentalists, such as *In nome del popolo inquinato* by Gianfranco Amendola (1987) and *La società dei rifiuti* by Giorgio Nebbia (1990). The year 1987 saw the publication of *Our Common Future*, the United Nations report also known as 'Brundtland Report', which defined sustainable development as 'development that meets the needs of the present without compromising the ability of future generations to meet their own needs'. The following years saw the launching of two periodic publications: the *Ambiente Italia* reports by the Istituto Ambiente Italia and the *Ecomafia* reports by Legambiente's Osservatorio Ambiente e Legalità. To date, both are still extraordinarily rich sources of data and information. Thanks to Giovanna Ricoveri's commitment, James O'Connor's eco-Marxist thinking took hold in Italy. O'Connor is the founder of the journal *Capitalism, Nature, Socialism*, which interprets the ecological crisis as a contradiction within capitalism.

New environmentalist associations sprung up alongside well-established ones like Legambiente, Italia Nostra, the WWF and the FAI. One was Slow Food, an association founded in 1986 in Bra, in the Province of Cuneo, through the activism of Carlo Petrini. Its purpose is to defend food biodiversity, as well as rural and food-and-wine traditions, against genetic manipulations and the homologation of food. Petrini was also the inspirer of the Università di Scienze Gastronomiche, the first and only university of taste in the world, founded in 2004 in Pollenzo in the province of Cuneo. Slow Food's activity has converged significantly with the work of Vandana Shiva, the founder of the Research Foundation for Science, Technology and Natural Resource Policy in India, and the promoter of many actions in defence of biodiversity and against the GMOs (Genetically Modified Organisms) introduced in India during the Green Revolution.

Several chairs of Ecology were founded in Italian universities, even in humanities faculties, as in the case of the chair of Political

Ecology held by Giorgio Nebbia at the University of Bari. Recognising the centrality of nature in social processes compelled scholars to address ecological issues from the perspectives of different disciplines, such as history, agronomy, town-planning studies, biology, economy, geology, sociology etc. In particular, in spite of strong resistance from mainstreamers, ecological economics started to take hold, which advocates an economic system based on renewable flows of energy and natural resources.

In the first fifteen years of the third millennium, Italian environmentalism penetrated deeply into the scientific and cultural debate, and into society; at the same time, it declined significantly in the political arena. While in the elections of 1992, 1994 and 1996 the Greens had retained shares between 2.5 and three per cent, in the subsequent ones they kept losing votes. Eventually, the introduction of an election threshold excluded them from Parliament, along with their extreme left-wing allies. Some exponents of the environmentalist world later joined the Ecodem group, the environmentalist area within the Partito Democratico. Few significant legislative measures were adopted during those years. The most noteworthy are the framework law on fires, approved in 2000; the introduction of the crime of illicit waste trade through the addition of Article 53 bis in the Ronchi Decree in 2001; the approval of a new charter for cultural heritage and the landscape known as 'Codice Urbani' in 2004; an act forbidding the mistreatment of animals passed in the same year; and the issuing of Legislative Decree 152, containing a revision of environmental norms, in 2006.

The Italian Green Party failed to develop a strong political structure, capable of translating the potential and originality it had displayed ever since the 1970s into stable consensus. According to many, the reason for this failure is that it was ridden with divisions and conflicts, and oscillated between movement and party (Alexander Langer called it an 'anti-party party'). According to others, Italian environmentalism was plagued by an excessively defensive and non-proactive attitude, which was branded as *ambientalismo del no* ('I-say-no' environmentalism). It had therefore failed to come up with a sweeping vision capable of situating environmental issues

within the context of an overall project to transform society and reduce inequalities. Furthermore, the party's transversal character and
the absence of true roots in an autonomous political tradition, as well
as the election of several important exponents of the 'green world' in
the ranks of the PCI (Partito Comunista Italiano; later Partito Democratico della Sinistra, Democratici di Sinistra, Ulivo and finally Partito Democratico), ultimately prevented it from forging a distinctive
and exclusive identity for itself. Others argue that the Green Party
failed to create stable collaborations with important forces such as
the feminist movement and historical associations like Italian Nostra, the WWF, Federnatura and the Amici della Terra. Others still
have pointed out that it failed to grasp the deep change the Italian
political system was going through in the early 1990s.

There is possibly a more general explanation – namely that the
crisis of political environmentalism is but a facet of what Alfio Mastropaolo called the 'major political depression' that has been affecting
Italian society for several decades. A result of the inadequacy of a
democracy to take on the new challenges posed by a rapidly changing society, this 'depression' has voided of their meaning concepts
such as 'public', 'collective' and 'general interest', so intimately welded
to a conception of politics founded on ideals of ecology and social
equity and on an idea of the state as the subject whose task it is to
pursue these ideals. Swept away by widespread hostility against the
traditional forms of politics, the objectives of the Italian green party
have been partly inherited by the fragmented universe of committees and movements that have arisen over the last fifteen years as a
response to real needs. These organisations, rooted in local areas and
no longer represented by traditional politics, include citizens' committees, landscape protection committees, zero-waste committees,
energy-waste committees, the No Tav, No Dal Molin, committees
against dumps and incinerators, antismog mothers, commons networks, public water committees, Terra dei Fuochi-Comitato per la
Salute and many others. Participants in this broad galaxy often adopt
generically anti-state stances, rather than showing a keen interest in
general and political solutions to problems.

3. Waste

The growth of waste production and the consequent difficulties in disposing of it belong to recent history. The beginnings of the phenomenon can be traced back to the 1950s, when an economy based on the consumption and destruction of commodities arose. Up to then, urbanisation and industry had reused materials in various ways.

The consequence of consumerism was the growth of waste. This waste was inadequately disposed of, mainly by removing it from the place where it was produced.

Since the 1970s, the national production of urban waste has almost trebled, from 13,000,000 to 33,000,000 tons in 2012. In the 1980s and 1990s, per capita average waste production grew at an unprecedented rate, from 248 kg in 1980 to 466 in 1998 – that is, the same rate as the aggregate and per capita GNP. After 1998, the per capita increase rate was lower. Annual production grew to 536 kg in 2012. More waste is produced in the central Italian Regions; here, in 2012, the average per inhabitant was 582 kg a year, more than in the North (503 kg) and the South (463 kg).

Special waste grew even more than urban waste. In 2012, 145,000,000 tons were disposed of. 91.8 per cent of this was non-dangerous, 8.2 dangerous. 84.6 per cent was recycled, 2.3 was processed to recover energy, 12.1 was dumped in landfills and one per cent incinerated. This waste includes a large variety of scraps coming from industries as well as organic and pharmaceutical chemical waste, waste of the healthcare sector and the photographic industry, depleted oils, paint, car tyres, solvents, asbestos, demolition debris etc.

Of course, the increased production of waste between the 1950s and 1970s was a consequence of the country's industrialisation. The reason for the near doubling of waste production in the 1980s, though, was the transformation of the distribution process. Advertising, marketing and the decline of unpacked goods in favour of branded products brought on a revolution in product packaging, spawning a whole new industry. Packaging has become a market strategy, determining a transformation and extraordinary increase in waste. In the United States, for example, packaging accounts for over a third of urban solid waste.

As to waste disposal, the situation within the Italian peninsula is extremely diversified. In the North-Centre, where an organisational fabric of public services and municipal companies exists, Lombardy has created an efficient system, although belatedly compared to other regions like Emilia-Romagna. In the South, the situation is different in regional capitals and in other towns. In general, however, there is a scarcity of industrial plants, compost plants and strong public companies. The prevalent disposal system is landfills, which are almost entirely in private hands.

As regards legislation, in 1997 the Ronchi Decree deeply reformed the waste sector. Its aim was to grant a high degree of protection of the environment and promote waste sorting and recycling. This decree was issued in the framework of a more general European policy which ever since the second half of the 1970s had sought to protect human beings and the environment from the noxious effects of waste and restore, save and protect natural resources. In line with these principles, the Ronchi decree sought to promote clean technologies, research on materials that would reduce the toxicity and volume of waste and the development of the technical capability to achieve these objectives. The whole management system provided for in the Ronchi decree was based on the reduction of disposal in landfills by increasing household waste sorting and using waste as fuel.

Over the last few decades, difficulties in disposing of waste have periodically resulted in serious emergencies, made evident by the accumulation of waste in city streets. One such emergency occurred in Florence in the 1980s. It was caused by the sudden decision to close the local waste-to-energy plant at a time when no nearby landfill offered a valid alternative. For a long time, Florence's waste was sent to Campania and Puglia. Another crisis occurred in Milan during the 1990s. It was determined by the saturation of authorised landfills and the impossibility of setting up new ones in the short term. In this case, too, for many months the waste of Milan was taken to other places. Palermo went through several waste emergencies, caused by financial crises arising from sleazy connections between the waste-management companies, the public administration and the Mafia.

The most serious waste crisis in Italy so far, however, was undoubtedly the one that hit Campania between the end of 2007 and the first few months of 2008, followed by several smaller ones between 2009 and 2010. Besides gaining worldwide media resonance, this emergency has left deep scars on the whole economic and manufacturing sector of the region. The emergency mainly concerned the metropolitan area of Naples. It was caused, concomitantly, by saturation of the region's landfills and a delay in the completion of the waste-to-energy plant at Acerra, whose construction had started in 2004 but was only concluded in 2009. Although the situation has improved since the plant opened, partly thanks also to an increase in waste sorting – especially in some areas in Campania – the situation remains precarious. About half of the waste produced must still be taken outside the region. The combustible part is carried by ship to the Netherlands, where it is processed in energy recovery plants. The organic part ends up in landfills in Puglia. This is a very onerous system whose costs are inevitably borne by the citizens.

The waste crises have exposed Italy's failure to set up disposal systems adequate to the growth of its waste production. Landfills are still the most widespread means of waste disposal. According to data regarding 2012, they receive 42 per cent of urban waste. There are, however, significant differences from one region to another. In Lombardy, for example, only a very small quota of waste ends up in landfills; 49.9 per cent is recycled and 44 per cent incinerated. Where processing plants are lacking or inadequate, the use of landfills peaks: 91 per cent in Sicily and Molise, 75 per cent in Calabria, 74 per cent in Liguria and 71 per cent in Lazio. It is worth stressing that data provided by the Istituto Superiore per la Protezione e la Ricerca Ambientale (Ispra) indicate that energy recovery has proved not to be a disincentive to recycling. The case of Lombardy is revealing in this regard; here energy is recovered from 31 per cent of waste while almost fifty per cent is recycled.

In Italy as a whole, the total percentage of waste sent to energy recovery plants is around nineteen per cent, which is low in comparison to European countries that are leaders in integrated waste processing. France has 128 waste-to-energy plants with an average processing capacity of 90,000 tons of waste annually. Germany has 94, but with an average capacity that is threefold that of the French plants. Italy has only 51, with an average capacity of 80,000 tons each.

The national average of waste recycling was around 42.3 per cent in 2013. This datum, however, conceals major local variations. From 2009 to 2013, recycling went up from 48 to 54.4 per cent in the North, from 24.9 to 36.3 per cent in the Centre and from 19.1 to 28.9 per cent in the South. The leading Regions were Trentino, Veneto, Friuli-Venezia Giulia and Piemonte with 64.6, 64.6, 59.1 and 54.6 per cent, respectively. The worst were Sicily, Calabria and Molise with 13.4, 14.7 and 19.9 per cent. Waste sorting has increased significantly in some regions, notably in Marche, from 29.7 per cent in 2009 to 60.27 in 2013, and in Campania, from 29.3 per cent to 44 per cent in the same period.

In the same year 2013, the leading Provinces were Treviso, with 78 per cent, Pordenone with 75 per cent and Belluno with 70.7 per cent. The Provinces with the lowest percentages were in Sicily, where Palermo, Messina, Enna and Siracusa did not exceed ten per cent. Among towns with more than 200,000 inhabitants, the best of 2013 were Verona with 45.9 per cent and Turin with 43.7. The worst were Rome with 29.1 per cent, Bari with 21.4 and Naples with 20.3.

There are various technical, cultural and economic factors holding back the growth of waste recycling. One is certainly cost. In Italy, the collection of household-sorted waste costs on average almost twice as much as that of unsorted waste, and even more if the waste is collected door-to-door. The profits obtained from the selling of recyclable materials cover only a low percentage of this cost. In any case, it is utopian to think that integral recourse to recycling can solve the waste problem. It remains an ex-post-action, which does not eliminate the causes of the processes that destroy natural resources. For the future, research is focusing on energy-saving production models and recyclable materials. Several industrial sectors going in that direction exist in the Netherlands, Denmark, Germany and Italy, too. Indeed, Italy is a European and world leader in the recycling of wood, glass, cardboard, aluminium and compost. The principle is that of minimising waste by employing production processes that do not involve the death of commodities, but plan them from the beginning to be recyclable as material for new commodities. This is what is recommended in the national waste prevention plan, issued in conformity with EC Directive 2008/98, the fountainhead of the modern discipline of urban waste management. Such a productive system would help to solve not only the waste problem, but also the problem of the consumption of non-reproducible resources.

4. 'Ecomafias'

According to Legambiente's *Rapporto ecomafia 2013*, 34,132 environmental crimes were reported in Italy in 2012, a 32.4 per cent increase compared to five years before. This kind of crime has become particularly widespread from the 1970s onward. It is largely concentrated in regions with a strong traditional Mafia presence, i.e., Campania, Calabria, Sicily and Puglia, but has extended from there to other Regions, not only in the South. Today we are confronted with what has come to be known as 'Ecomafia'.

The toxic waste trade is the most deeply-rooted Ecomafia activity in Italy, and the one that has attracted the most media attention over the last few years. Metal works, paper mills and tanning industries, prevalently located in Piemonte, Lombardy, Emilia-Romagna, Veneto and Liguria, have often entrusted – through intermediaries – their waste disposal to criminal organisations. They have thus fuelled a huge southward traffic of cadmium, zinc, paint waste, steelworks slag, lead, asbestos, colorants, by-products of the aluminum industry, etc. This phenomenon was encouraged by the backwardness of the Italian waste recovery, reuse and incineration system, compared to which disposal in a landfill, even an illegal one, was easier and a lot cheaper. According to the Cassiopea investigation – the first to deal with the illicit waste trade between northern Italy and Campania in the 1980s and 1990s – illegal disposal cost from six to fifteen times less than legal disposal: eight to 200 Lire per kilogram instead of 800–1,200. Illicit disposal was thus both a cheap solution for industries and a lucrative venture for criminal organisation; for a long time, it was also a low-risk activity, since illicit waste trafficking was only established as a crime in 2001, in Article 260 of the environment code. Previously it had only been sanctioned with modest fines.

Chapter Four

Figure 4.4

Waste crisis, Caserta, southern Italy. Source: Wikimedia commons

To quote Nunzio Perrella, a Mafia affiliate turned informant, 'trash is gold' ('*a munnezza è oro*).

Some areas in Campania in the Provinces of Caserta and Naples were regarded as Ecomafia territory par excellence. These are the areas where the Casalesi clan gained sway by reusing quarries as illegal landfills, and thus managed to integrate the cement cycle and the waste disposal cycle. The clans brought to the waste traffic the know-how and resources – places, people, equipment – they had acquired in the production of inert materials and concrete, on which they had had a monopoly ever since the first half of the 1980s. Once a concrete quarry was exhausted, it was reused as a landfill. Many other criminal groups beside the Casalesis were involved both in the waste and in the concrete trade.

The pollution of the soil and the water caused by the illicit disposal of industrial waste has been compounded by air pollution caused by its open-air combustion. This practice, meant to erase all traces of the origin of waste, has become widespread in a vast area between Naples and Caserta, hence dubbed 'Terra dei Fuochi' (Land of Fires). The Terra dei Fuochi does not owe its notoriety merely to its pollution issues – which actually only concern limited areas – but

Figure 4.5

Illegal building in central Italy.
Source: http://www.monitorimmobiliare.it/abusivismo-edilizio-il-picco-sulle-coste-del-lazio-

also to the way in which the industrial waste traffic is entangled with the district's more general process of landscape and social deterioration. Illegality and illicit building have put the stamp on the development of the area. It is ridden with housing lots that are often unconnected to services and infrastructure, scraps of city lacking squares or sidewalks, tatters of abandoned farmland, open-air sewers, mounds of abandoned waste and evil-smelling dumps.

In the other regions with a long tradition of Mafia presence, criminal businesses have also turned to the waste disposal sector. Puglia has become a stopover for industrial waste being shipped abroad, prevalently from the ports of Bari and Taranto. This trade is mainly managed by non-local organisations, but the local Mafia – the Sacra Corona Unita – also has a hand in it. In Calabria, the *'ndrangheta* manages the industrial waste traffic from outside the region, which has been growing ever since 2000. In Sicily, as early as the 1990s Mafia organisations have prevalently focused on local waste, controlling tenders for the building and management of landfills and for waste collection and transport.

Organised crime is also involved in the illegal waste trade in Piemonte, Liguria, Lombardy and Veneto. It is from these regions that special waste heads south, and even towards other northern regions, notably from Lombardy to Piemonte, Veneto and Emilia-Romagna, where stocking sites were found whence industrial waste was shipped to the South. The Mafias have been infiltrating the local economies of north-central Italian regions ever since the 1980s. The Abruzzo Region has played an important role as a stopover along the Adriatic route. Latium is also an important transit area for the waste trade, and especially the province of Frosinone, since it adjoins the province of Caserta. It is also worth noting that the waste traffic extends far beyond the national boundaries, even to remote countries. According to data from 2012, Italian investigations for waste trafficking crimes involved 26 countries – eleven in Europe, six in Asia and eight in Africa. The largest-scale traffics are those of car tyres towards South Korea, India and Thailand, and of metal towards China.

Another major sector that has attracted the interest of criminal organisations is construction. Here illegality manifests itself in various ways, including illegal building in areas lacking town plans, the trumping of tenders, unauthorised building of housing lots, unauthorised increases of building volumes, granting building permits that violate planning rules and entrusting building works to companies with ties to criminal groups. Illegal construction is not a problem only in regions with long-running Mafia traditions, but also in Liguria, Emilia-Romagna and Tuscany, and is especially rampant along the seaside and in scenic landscape areas. According to an investigation by the CRESME (Centro Ricerche Economiche, Sociali, di Mercato per l'Edilizia e il Territorio), in the South the illegality mostly takes the form of ex novo building, whereas in the North it mostly concerns restorations and extensive renovations. Illegal construction causes several kinds of environmental damage, from hydrogeological risk – when buildings are erected in unsuitable places, in absence or neglect of town-planning standards and with inadequate infrastructure (aqueducts, communication systems, sewers) – to the defilement of valuable landscapes.

Environment and environmentalism in contemporary Italy

Ecomafias could never have flourished without the close-knit web of ties they have woven in public administrations – implicating high officials, mayors, city councilors and heads of planning offices – and in the world of professions, companies and banks, according to various degrees of connivance, ranging from complicity to corruption and outright Mafia membership. In the 1990s, many town governments were dissolved due to Mafia infiltration related to criminal activities in the waste and concrete sectors.

These *alleanze nell'ombra* (alliances in the shade), to use Rocco Sciarrone's definition, have enabled criminal organisations to rig public tenders by means of frauds or threats; to obtain waste disposal permits without the required authorisations and make sure that the authorities omit to verify forged certificates and delivery notes for toxic waste; and to demolish illegal buildings. The absence of governmental control has often caused construction work to be assigned to companies lacking the necessary qualifications, but sponsored by criminal groups. A criminal business may camouflage itself behind a number of small companies it controls, or use different subjects to provide a service, or purchase branches of 'clean' companies. Once more, the weakness of public administrations – which often comprise accomplices of the clans, systematically fail to control their territory and, above all, lack bureaus and services adequate to this task – re-emerges as a distinctive trait of environmental history in Italy. The *Ecomafia* report of 2013 indicates that the highest number of legal investigations of environmental corruption in 2012 was conducted in Lombardy, 14.8 per cent of the total. The highest number of arrests for crimes against the environment was made in Calabria, followed by Campania, Lombardy, Tuscany and Piemonte.

Besides waste disposal, concrete and incendiarism, Ecomafias are also involved in other sectors. In agriculture, they control cultivation and harvests by exploiting prevalently immigrant labour and imposing their products over the whole supply chain. There are also criminal infiltrations in renewable energy sectors benefiting from special public funding, as was the case with wind turbines in Sicily. Criminal organisations are also undermining Italy's artistic and cultural heritage. In 2012, the revenue from the theft and illegal

trading of artworks and archaeological artifacts was higher than €
267,000,000. Finally, the animal racket encompasses the hide and fur
industry and the trade of rare species (turtles, snakes, tigers, leopards,
rhinoceroses and elephants). This trade extends across three continents – Europe, Asia and Africa – and, like the waste traffic, bears
witness to the global character Ecomafias have taken on today. Sicily
is the region with the highest number of recorded violations in this
sector in 2012, followed by Puglia, Campania and Lazio.

It is important to stress that one of the reasons that criminal businesses have prospered in environment-related sectors is the
way Italians have looked at the relationship between the environment and economic growth, and the way in which they have managed development, especially after World War II. Italy has neglected
to defend its historic and natural identity. Its public planning has
been weak, allowing owners' rights to prevail over the conservation
of environmental resources. Its local administrations have proved inadequate and have shown a serious lack of ecological culture. The
result, as we have seen, has been a chaotic urbanisation which has
produced uncontrolled land consumption, devastation of fertile land
and scenic landscapes, pollution, sanitary problems, inadequate public transportation and social distress. In this context, aggravated by
an absence of habitat defence policies ever since the beginning of
the industrialisation process, the waste problem has grown, and the
country has failed to respond by implementing adequate disposal
and processing systems. All this has created dangerous openings for
criminal infiltration. Environmental crime is thus not a mere public
order problem. It cannot be effectively eradicated without a general
reconsideration, at once cultural and political, of our relationship
with the environment.

CONCLUSIONS

The environmental situation of present-day Italy is diverse and contradictory. On the one hand, public policies have proved incapable of curbing land consumption and illegal building, keeping in check hydrogeological risk, creating complete cycles for the disposal of urban and industrial waste, fighting environmental crime, promoting more efficient public transportation systems and restoring high-risk areas. At the same time, in some departments Italy is as advanced as Germany, a politically and economically 'green' country. Italy has reduced its natural resource and energy consumption, as well as its carbon dioxide emissions. This is what Duccio Bianchi called an 'unconscious' transformation of some sectors of the economy. Italy has shown an inclination to green economy – that is, to a low environmental impact economy. Over the last few years, renewable energies – wind turbines, solar energy, biomasses, hydroelectricity, geothermal energy – have grown significantly, accounting today for almost thirty per cent of the national electric energy requirement (the leading Regions in this sector are Lombardy and Puglia). Italy also ranks among the ten countries in the world with the highest percentage of organic farmland. Over forty Italian towns house within their limits more than a hundred organic companies. Noto in the province of Syracuse has 446, Corigliano Calabro in the Province of Cosenza 242 and Poggio Moiano in the Province of Rieti 241.

It is Italy's great variety of agriculture and land – which depend not only on the gifts that nature bestowed on the Belpaese in the first place, but also on the long and rich history of the country's rural landscapes – that constitute its 'sustainable' economic resource, which public policies have often failed to adequately recognise and promote. The country's extraordinary variety of food specialties (4,698 have been counted) and its great regional culinary traditions make Italy one of the leading countries in wine and food tourism. According to data of Coldiretti, it has achieved worldwide primacy in this sector, thanks to its range of high-quality products: from wine to pasta, from seafood to cured meats. Finally, Italy is a leader in the industry of recycling, particularly of metal, plastic and textiles, but also of paper, glass, wood and rubber. In 2010, it was the country in Europe that recycled the highest quantity of material.

These of course are only encouraging signs, but they do suggest that Italy may eventually manage to convert its economic and energy system to more sustainable forms of production. However, in spite of the successes green economy has been scoring lately, the idea that it could provide a pathway for Italy's economic recovery holds a very marginal position in the national government's political agenda. There is strong resistance against the green economy not only in public administrations, but also in the principal ministries, which do not seem to look favourably on a development model founded on the principle of sustainable economy. To implement such a model, we would need to totally redesign the technology of whole sectors, such as the automobile industry, steering them towards zero emission production lines. The generalised employment of recyclable material in the industry should be favoured in order to reduce waste production. Sustainable transportation systems should be expanded and urban requalification policies promoted. Urban soil consumption needs to be kept in check. Investments and employments in the so-called 'green sectors' should be promoted. This list of 'to dos' could go on. These changes, however, clash with strong interests, not only those of major economic actors but often also those of local populations, which are themselves responsible for forms of environmental destruction, as in the case of illegal construction. All this is connected to the character of Italy's democratic system, its mechanisms of political representation and the quality of the political class these mechanisms select. Today the dilemma is how to conjugate environmental protection and production of wealth, jobs for everyone and political consensus. Education undoubtedly plays a role in this. People need to be informed about how ecosystems work, about the serious social consequences of their disruption, and about the economic benefits of their conservation and protection. The political decisions that will, or will not, be made on the environment are destined to have an effect on the future life, wealth-producing ability and well-being of Italian citizens today and tomorrow.

ACKNOWLEDGEMENTS

This book is meant as a handy summarisation of a long study and research journey. It has benefited from the contributions and support of male and female scholars from several disciplines. I thank all of them, but particularly the environmental historians among them, whose interests have been closest to my own. I am especially grateful to Paolo Malanima and Simone Neri Serneri, who have been important interlocutors during this research journey and helped me to bring this book to completion. Special, deep and heartfelt thanks go to Piero Bevilacqua, from whom I learned much of what is written in the following pages, and who pointed out and illuminated the path to follow.

The Author

Gabriella Corona is a senior researcher at the Institute of Studies on Mediterranean Societies of the National Research Council in Naples and teaches Economic History in the Humanities Department of University of Naples 'Federico II'. She is co-director of *Meridiana. Rivista di storia e scienze sociali* and co-editor in chief of *Global Environment: A Journal of Transdisciplinary History*. Her books include *Demani ed individualismo agrario nel Regno di Napoli* (1995); *I ragazzi del piano. Napoli e le ragioni dell'ambientalismo urbano* (2007); *Rifiuti. Una questione non risolta* (2010), with D. Fortini; *The Problem of Waste Disposal in a Large European City* (2012), with D. Fortini; and the Italian edition of the present work, *Breve Storia dell'Ambiente in Italia* (2015). She has published many environmental history essays and has edited books including *Storia e Ambiente* with S. Neri Serneri (2007) and *Economia e ambiente in Italia dall'Unità a oggi* (2012), with P. Malanima.

BIBLIOGRAPHY

Chapter One

1.

Among general works, the following will suffice: A. Caracciolo, *L'ambiente come storia. Sondaggi e proposte di storiografia dell'ambiente*, Bologna: il Mulino, 1988; C. Ponting, *A Green History of the World*, New York: Penguin Books, 1991; A. Crosby, *Ecological Imperialism. The Biological Expansion of Europe 900-1900*, Cambridge: Cambridge University Press, 1993; J.R. McNeill, *Something New Under the Sun. An Environmental History of the Twentieth-Century World*, New York: Norton, 2000; K. Pomeranz, *The Great Divergence: China, Europe and the Making of the Modern World Economy*, Princeton, NJ: Princeton University Press, 2000; P. Bevilacqua, *La terra è finita. Breve storia dell'ambiente*, Roma-Bari: Laterza, 2006.

On changes in European agriculture: J.A. Chambers and G.E. Mingay, *The Agricultural Revolution 1750-1880*, London, Batsford: 1966; B.H. Slicher Van Bath, *The Agrarian History of Western Europe. A.D. 500-1850*, London: Edward Arnold, 1966; D. Grigg, *Population Growth and Agrarian Change. An Historical Perspective*, Cambridge: Cambridge University Press, 1980. The reflections of E.A. Wrigley, *Continuity, Chance and Chanmge. The Character of the Industrial Revolution in England*, Cambridge: Cambridge University Press, 1990, are also important.

On the energy transition: P. Malanima, *Energia e crescita nell'Europa preindustriale*, Roma: la Nuova italia Scientifica, 1996; id., 'Tra due sistemi energetici. I consumi di energia in Europa tra il 1600 e il 1800', *Meridiana* **30** (1997): 17–39; id., *Uomini, risorse, tecniche nell'economia europea dal X al XIX secolo*, Milano: Bruno Mondadori, 2003; id., *Le energie degli italiani. Due secoli di storia*, Milano: Bruno Mondadori, 2013.

On the decline of common land restrictions: D. Demélas and N. Vivier (eds), *Les propriétés collectives face aux attaques libérales (1750-1914). Europe occidentale et Amérique latine*, Rennes: Presses Universitaires de Rennes, 2003. A very interesting study on the role of the peasant economy and communal systems in the agrarian revolution is

Gabriella Corona

R.C. Allen, *Enclosure and the Yeoman. The Agricultural Development of the South Midlands 1450-1850*, Oxford, New York: Clarendon Press, 1992. See also M. Bloch's classic study *Les caractéres originaux de l'histoire rurale française*, Paris: Colin, 1988.

2.

On the original features of the Italian landscape, L. Gambi, 'I valori storici dei quadri ambientali', G. Haussmann, 'Il suolo d'Italia nella storia', E. Sereni, 'Agricoltura e mondo rurale', all in *Storia d'Italia, i: I caratteri originali*, Torino: Einaudi, 1972, respectively pp. 5–60, pp. 63–132 and pp. 135–252; P. Bevilacqua, *Tra natura e storia. Ambiente, economia, risorse in Italia*, Roma: Donzelli, 1996; id., 'I caratteri originali della storia ambientale italiana', *I frutti di Demetra* **8** (2005): 5–13; F. Cazzola, 'Sui caratteri originali della storia ambientale italiana', *I frutti di Demetra* **9** (2006): 5–12.

On the ancient roots of the transformation of the farming landscape in Italy, D. Moreno, *Dal documento al terreno. Storia e archeologia dei sistemi agro-silvo-pastorali*, Bologna: il Mulino, 1990; P. Bevilacqua, 'Linee generali per la costruzione di un Catalogo del paesaggio agrario italiano', *I frutti di Demetra* **14** (2007): 5–15. See also M. Agnoletti (ed.), *Paesaggi rurali storici. Per un catalogo nazionale*, Roma/Bari: Laterza, 2010.

For an analysis of demographic trends in the pre-unification period and their effects on ecological equilibria, P. Malanima, *La fine del primato. Crisi e riconversione nell'Italia del Seicento*, Milano: Bruno Mondadori, 1998.

On the hydrogeological fragility and geological features of the Italian peninsula, a fundamental text is G. Gisotti and M. Benedini, *Il dissesto idrogeologico. Previsione, prevenzione e mitigazione del rischio*, Roma: Carocci, 2000. For a more strictly historical analysis of the same phenomenon, P. Bevilacqua, 'Catastrofi, continuità, rotture nella storia del Mezzogiorno', *Laboratorio politico* **5–6** (1981): 177–219; W. Palmieri, 'Le catastrofi rimosse: per una storia delle frane e delle alluvioni del Mezzogiorno continentale', *Meridiana* **44** (2002) 97–124; id., 'Dissesto e disastri idrogeologici nell'Italia unita', in G. Corona and P. Malanima (eds), *Economia e ambiente in Italia dall'Unità a oggi*, Milano: Bruno Mondadori, 2012, pp. 125–145.

Bibliography

On woods in the pre- and post-unification periods, R. Sansa, 'Il bosco fra difesa degli usi consuetudinari e conflitto di mercato', *Storia urbana* **69** (1994): 133–149; id., 'Usi del bosco: modalità di attivazione delle risorse a confronto', *Storia urbana* **76–77** (1996): 203–212; M. Armiero, *Il territorio come risorsa. Comunità, economie e istituzioni tra i boschi meridionali*, Napoli: Liguori, 1999; W. Palmieri, 'Il bosco nel Mezzogiorno preunitario tra legislazione e dibattito' and R. Sansa, 'Il mercato e la legge: la legislazione forestale italiana nei secoli XVIII e XIX', both in P. Bevilacqua and G. Corona (eds), *Ambiente e risorse nel Mezzogiorno contemporaneo*, Corigliano Calabro: Meridiana libri, 2000, pp. 27–62 and 3–26. On changes in the use of woodland after unification, M. Agnoletti, 'Osservazioni sulle dinamiche dei boschi e del paesaggio forestale italiano fra il 1862 e la fine del secolo XX', in *Società e storia* (2005): 377–396; id., 'Foreste e paesaggio', in Corona and Malanima (eds), *Economia e ambiente in Italia dall'Unità a oggi*, cit., pp. 99–123.

On land reclamation in Italian history, P. Bevilacqua and M. Rossi-Doria, *Le bonifiche in Italia dal '700 ad oggi*, Roma, Bari: Laterza, 1984. On land reclamation under the Bourbon monarchy, P. Bevilacqua, *Breve storia dell'Italia meridionale dall'Unità a oggi*, Roma: Donzelli, 1993, especially pp. 9–15; on legislation against hemp retting, C. D'Elia, 'Uso delle risorse e tentativi di riforma: la macerazione di canapa e lino nel primo Ottocento', in Bevilacqua and Corona (eds), *Ambiente e risorse nel Mezzogiorno contemporaneo*, cit., pp. 157–166. On deforestation in the Italian South and its effects on the relationship between the mountains and the plains, P. Tino, 'La montagna meridionale. Boschi, uomini, economie tra Ottocento e Novecento', in P. Bevilacqua (ed.), *Storia dell'agricoltura italiana in età contemporanea, I: Spazi e paesaggi*, Venezia: Marsilio, 1989, pp. 677-754; id., *Le radici della vita. Storia della fertilità della terra nel Mezzogiorno (secoli XIX-XX)*, Roma: Xl edizioni, 2010.

On the history of fishing, M. Gangemi, 'Pesce, spugne e coralli. La Grande pesca italiana dal Mediterraneo all'Atlantico (1879-1938)', in V. D'Arienzo and B. Di Salvia (eds), *Pesci, barche, pescatori nell'area mediterranea dal medioevo all'età contemporanea*, Milano: Angeli, 2010, pp. 138–184; for an environmental historical approach to the same theme, M. Armiero, 'L'Italia di Padron 'Ntoni. Pescatori, legislatori e burocrati tra XIX e XX secolo', in P. Frascani (ed.), *A vela e a vapore. Economie, culture e istituzioni del mare nell'Italia dell'Ottocento*, Roma:

Donzelli, 2001, pp. 177–213. An interesting case study is A. Clemente, *Il mestiere dell'incertezza. La pesca nel Golfo di Napoli tra XVIII e XX secolo*, Napoli: Guida, 2005.

On the history of earthquakes, besides the fundamental study by E. Boschi, E. Guidoboni, G. Ferrari, G. Valensise and P. Gasperini, *Catalogo dei forti terremoti in Italia dal 461 al 1990*, Bologna: Istituto Nazionale di Geofisica-Sga, 1997, see E. Guidoboni, 'Un'antirisorsa del Sud: i disastri sismici nella sfida economica', in Bevilacqua and Corona (eds), *Ambiente e risorse nel Mezzogiorno contemporaneo*, cit., pp. 245–261; id., 'Dimenticare i terremoti? I segni dell'attività sismica nel paesaggio culturale e naturale in Italia', in P. Bevilacqua and P. Tino (eds), *Natura e società. Studi in memoria di Augusto Placanica*, Roma: Donzelli, 2005, pp. 17–36. See also G. Parrinello, *Fault Lines: Earthquakes and Urbanism in Modern Italy*, Oxford, New York: Berghahn Books, 2015.

3.

On the role of natural resources and environmental factors in the various facets of the organisation of production in Italy, *Bevilacqua, Tra natura e storia. Ambiente, economie, risorse in Italia*, cit.

For this chapter, I drew mainly on my essay 'Territorio produttivo e modelli di sviluppo. I contributi della ricerca recente', *Meridiana* 30 (1997): 107–133. On the history of the *cascina*, F. Cazzola, *Storia delle campagne padane dall'Ottocento a oggi*, Milano: Bruno Mondadori, 1996; G. Crainz, 'La cascina padana. Ragioni funzionali e svolgimenti', in Bevilacqua (ed.), *Storia dell'agricoltura italiana in età contemporanea*, I, cit., pp. 37–76; B. Bianchi, 'La nuova pianura. Il paesaggio delle terre bonificate in area padana', in Bevilacqua (ed.), *Storia dell'agricoltura italiana in età contemporanea*, I, cit., pp. 451–494; G. Della Valentina, 'Padroni, imprenditori, salariati: modelli capitalistici padani', in P. Bevilacqua (ed.), *Storia dell'agricoltura italiana in età contemporanea, II: Uomini e classi*, Venezia: Marsilio, 1990, pp. 151–200.

On the *fattoria* and its evolution, G. Giorgetti, *Contadini e proprietari nell'Italia moderna. Rapporti di produzione e contratti agrari dal secolo XVI a oggi*, Torino: Einaudi, 1974; L. Bellicini, 'La campagna urbanizzata. Fattorie e case coloniche nell'Italia centrale e nord-orientale', in Bevilacqua (ed.), *Storia dell'agricoltura italiana in età contemporanea*, I, cit., pp. 77–130; S. Anselmi, 'Mezzadri e mezzadrie nell'Italia centrale', in Bevilacqua (ed.),

Bibliography

Storia dell'agricoltura italiana in età contemporanea, II, cit., pp. 201–220; G. Biagioli, 'Il podere e la piazza. Gli spazi del mercato agricolo nell'Italia centro-settentrionale', in P. Bevilacqua (ed.), *Storia dell'agricoltura italiana in età contemporanea, III: Mercati e istituzioni*, Venezia: Marsilio, 1991, pp. 3–63; G. Biagioli and R. Pazzagli (eds), *Mezzadri e mezzadrie tra Toscana e Mediterraneo*, Pisa: Felici, 2013.

On agriculture without houses, C. Maranelli, *Considerazioni geografiche sulla questione meridionale*, Bari: Laterza, 1946; M. Rossi Doria, 'Il Mezzogiorno agricolo e il suo avvenire: "l'osso e la polpa"', in *Nord e Sud nella società e nell'economia italiana d'oggi*, Torino: Proceedings of conference organised by Fondazione Luigi Einaudi, 1968, pp. 285–299; P. Bevilacqua, 'Uomini, terre, economie', in *Storia d'Italia. Le regioni dall'Unità ad oggi. La Calabria*, Torino: Einaudi, 1985, pp. 117–362; F. Mercurio, 'Agricolture senza casa. Il sistema del lavoro migrante nelle maremme e nel latifondo', in Bevilacqua (ed.), *Storia dell'agricoltura italiana in età contemporanea*, I, cit., pp. 131–179; A. Massafra and S. Russo, 'Microfondi e borghi rurali nel Mezzogiorno', in Bevilacqua (ed.), *Storia dell'agricoltura italiana in età contemporanea*, I, cit., pp. 181–228; E. Sori, 'Popolazioni e insediamenti nel Mezzogiorno contemporaneo', *Meridiana* **10** (1990): 46–76.

On the role of water in the transformation of southern farming economies, P. Bevilacqua, *Le rivoluzioni dell'acqua. Irrigazioni e trasformazioni dell'agricoltura tra Sette e Novecento*, in id. (ed.), *Storia dell'agricoltura italiana in età contemporanea*, I, cit., pp. 255–318; S. Lupo, *Il giardino degli aranci. Il mondo degli agrumi del Mezzogiorno*, Venezia: Marsilio, 1990; P. Bevilacqua, 'L'acqua e le trasformazioni ambientali nel Sud moderno e contemporaneo', in V. Teti (ed.), *Storia dell'acqua. Mondi materiali e universi simbolici*, Roma: Donzelli, 2003, pp. 129–136.

On the history of technical innovations in agriculture in the nineteenth and twentieth centuries, G. Corona and G. Massullo, 'La terra e le tecniche. Innovazioni produttive e lavoro agricolo nei secoli XIX e XX', in Bevilacqua (ed.), *Storia dell'agricoltura italiana in età contemporanea*, I, cit., pp. 353–449.

On emigration', P. Bevilacqua, A. De Clementi and E. Franzina (eds), *Storia dell'emigrazione italiana*, Roma: Donzelli, 2001–2002.

Gabriella Corona

Chapter Two

1.

Among the main Italian studies in the recent debate on commons, see especially U. Mattei, *Beni comuni. Un manifesto,* Roma, Bari: Laterza, 2011; P. Cacciari (ed.), *La società dei beni comuni,* Roma: Ediesse, 2011; A. Ciervo, *I beni comuni,* Roma: Ediesse, 2012; G. Arena and C. Iaione (eds), *L'Italia dei beni comuni,* Roma: Carocci, 2012; L. Pennacchi, *Filosofia dei beni comuni. Crisi e primato della sfera pubblica,* Roma: Donzelli, 2012; Fondazione Lelio e Lisli Basso-Issoco, *Tempo di beni comuni. Studi multidisciplinari,* Roma: Ediesse, 2013; E. Vitale, *Contro i beni comuni. Una critica illuminista,* Roma, Bari: Laterza, 2013.

Outside of Italy, the literature on the subject is vast. Two studies, in particular, have become classics: G. Hardin, 'The Tragedy of the Commons', *Science* December 1968: 1243–1248; E. Ostrom, *Governing the Commons. The Evolution of Institutions for Collective Action,* Cambridge: Cambridge University Press, 1990, reviewed in G. Corona, 'Diritto e natura: la fine di un millennio', *Meridiana* 28 (1997): 127–161.

For this section, I drew on three essays of mine, 'Declino dei "commons" ed equilibri ambientali. Il caso italiano tra Otto e Novecento', *Società e storia* **104** (2004): 357–383; 'Paolo Grossi e la risposta italiana alla "Tragedy of the Commons"', *I frutti di Demetra* **1** (2004): 9–15; 'La questione dei beni comuni in Italia', *Proposte e ricerche* **71** (2013): 168–183. On the history of collective property and common rights, P. Grossi, *Un altro modo di possedere. L'emersione di forme alternative di proprietà alla coscienza giuridica post-unitaria,* Milano: Giuffrè, 1977; M. Caffiero, *L'erba dei poveri, comunità rurale e soppressione degli usi collettivi nel Lazio (secoli XVIII-XIX),* Roma: Ed. dell'Ateneo, 1983; R. Ago, 'Conflitti e politica del feudo: le campagne romane del Settecento' and E. Grendi, 'La pratica dei confini: Moglia e Sassello, 1715-1745', both in *Quaderni storici* **63** (1986); B. Farolfi, *L'uso e il mercimonio. Comunità e beni comunali nella montagna bolognese del settecento,* Bologna: Clueb, 1987; G.C. De Martin (ed.), *Comunità di villaggio e proprietà collettive in Italia e in Europa,* Padova: Cedam, 1990; the acts of the conference *Terre e comunità nell'Italia Padana. Il caso delle Partecipanze agrarie emiliane: da beni comunali a beni collettivi .* Proceedings of the conference, Nonantola, 16–18 Nov. 1990, published in *Cheiron* **VIII** 14–15 (1990–

123

Bibliography

1991): 33–100; O. Raggio, 'Forme e politiche di appropriazione delle risorse: casi di usurpazione delle comunaglie in Liguria', *Quaderni storici* 79 (1992) 135–170; volume 81 (1992) of *Quaderni storici*, devoted to collective resources; G. Corona, *Demani ed individualismo agrario nel Regno di Napoli (1780-1806)*, Napoli: Esi, 1995; P. Nervi (ed.), *I demani civici e le proprietà collettive. Un diverso modo di possedere. Un diverso modo di gestire*, Padova: Cedam, 1998.

On the juridical-historical history of the nineteenth and twentieth century, G. Curis, *Usi civici, proprietà collettive e latifondo nell'Italia centrale e nell'Emilia con riferimento ai demani comunali del Mezzogiorno*, Napoli: Jovene, 1917; id., *L'evoluzione degli usi civici delle ex provincie pontificie*, Roma: Tip. dell'Unione cooperative Editrice, 1907; F. Lauria, *Demani e feudi nell'Italia meridionale*, Napoli: Tip. degli Artigianelli, 1923; M. Palumbo, *I comuni meridionali prima e dopo le leggi eversive della feudalità*, 2 vols., Cerignola: l'Unione, 1910–1916; R. Trifone, *Gli usi civici*, Milano: Giuffrè, 1963, id., *Feudi e demani. Eversione delle feudalità nelle provincie meridionali. Dottrina, storia, legislazione e giurisprudenza*, Milano: Società Editrice Libraria, 1909; A. Cencelli-Perti, *La proprietà collettiva in Italia. Le origini, gli avanzi, l'avvenire*, Roma: Manzoni, 1890; V. Lombardi, *Delle origini e delle vicende degli usi civici nelle provincie napoletane. Studio storico-legale*, Cosenza: Tip. Municipale, 1882; *Atti della Commissione Reale pei demani comunali nelle provincie del Mezzogiorno istituita con R. Decreto 4 maggio 1884*, Roma: Bertero and C., 1902; G. Raffaglio, *Diritti promiscui, demani comunali, usi civici*, Milano: Società Editrice libraria, 1915. Ghino Valenti's reflections referred to in the text, in *Atti della giunta per la inchiesta agraria e sulle condizioni della classe agricola*, Roma: Forzani, 1883, XI/2.

On the role of land ownership in moderate political planning in the Risorgimento, A.M. Banti, 'I proprietari terrieri nell'Italia centro-settentrionale' and S. Lupo, 'I proprietari terrieri nel Mezzogiorno', both in P. Bevilacqua (ed.), *Storia dell'agricoltura italiana in età contemporanea, II: Uomini e classi*, Venezia: Marsilio, 1990, pp. 43–103 and 105–149.

On common land rights and collective ownership in Italy after World War II, see G. Medici, *La distribuzione della proprietà fondiaria in Italia. Relazione generale*, Roma: Edizioni italiane, 1948; C. Federico

Gabriella Corona

and R. Giacoia, 'Verso un accesso facilitato agli indici del Bollettino degli usi civici', *Archivio Scialoja-Bolla. Annali di studi della proprietà collettiva* 1 (2004): 145–163.

2.

On the history of relations between cities and the environment, E. Sori, *La città e i rifiuti. Ecologia urbana dal Medioevo al primo Novecento*, Bologna: il Mulino, 2001; S. Neri Serneri, *Incorporare la natura. Storie ambientali del Novecento*, Roma: Carocci, 2005; G. Corona, 'Ecosistema città', in G. Corona and P. Malanima (eds), *Economia e ambiente in Italia dall'Unità a oggi*, Milano: Bruno Mondadori, 2012, pp. 9–30. On the case studies, see G. Corona and S. Neri Serneri (eds), *Storia e ambiente. Città, risorse e territori nell'Italia contemporanea*, Roma: Carocci, 2007, especially the essays by M. Nucifora, M.L. Ferrari, M. Morgante, A. Ciuffetti, R. Parisi, O. Aristone, A.L. Palazzo, V. Bulgarelli and C. Mazzeri.

On the role of technology in the transformation of natural resources in urban contexts, stressed by American urban environmental history ever since the first half of the 1980s, see M. Melosi's classic study *The Sanitary City*, Baltimore/London: The Johns Hopkins University Press, 2000.

For statistical data on population, P. Malanima, 'Urbanisation and the Italian economy during the last millennium', *European Review of Economic History* 9/1 (2005): 97–122; Svimez, *150 anni di statistiche italiane: Nord e Sud, 1861-2011*, Bologna: il Mulino, 2011.

On the building of aqueducts, P. Celentani Ungaro, 'L'opera della "Cassa" per gli acquedotti e le fognature', in Centro studi Cassa per il Mezzogiorno (eds), *Cassa per il Mezzogiorno. Dodici anni 1950-1962, III.1: Acquedotti e fognature*, Bari: Laterza, 1963; G. Bigatti, 'La conquista dell'acqua. Urbanizzazione e approvvigionamento idrico', in G. Gigatti, A. Giuntini, A. Mantegazza and C. Rotondi, *L'acqua e il gas in Italia. La storia dei servizi a rete, delle aziende pubbliche e della Federgasacqua*, Milano: Angeli, 1997, pp. 27–161; E. D'Angelis and A. Irace, *Il valore dell'acqua. Chi la gestisce, quanta ne consumiamo e come possiamo salvarla*, Milano: Dalai, 2011; A. Massarutto, *Privati dell'acqua? Tra bene comune e mercato*, Bologna: il Mulino, 2011.

On the hygienic revolution in Italy, C. Giovannini, *Risanare le città. L'utopia igienista di fine Ottocento*, Milano: Angeli, 1996; C. Pogliano,

Bibliography

'L'utopia igienista (1870–1920)', in F. Della Peruta (ed.), *Storia d'Italia. Annali, VII: Malattia e medicina,* Torino: Einaudi, 1997, pp. 589–601; G. Zucconi, *La città contesa. Dagli ingegneri sanitari agli urbanisti (1855–1942),* Milano: Jaca book, 1999.

On the history of pollution in Europe, a classic is C. Bernhardt and G. Massard-Guilbaud (eds), *Le démon moderne. La pollution dans les sociétés urbaines et industrielles d'Europe/The Modern Demon. Pollution in Urban and Industrial European Societies,* Clermont-Ferrand: Presses Universitaires Blaise Pascal, 2005.

On cultural changes in the relationship between human beings and nature, A. Corbin, *Le territoire du vide. L'Occident et le désir du rivage (1750–1840),* Paris: Flammarion, 1988; K. Thomas, *Man and the Natural World. Changing Attitudes in England 1500–1800,* London : Allen Lane, 1983; P. Sorcinelli, *Storia sociale dell'acqua. Riti e culture,* Milano: Bruno Mondadori, 1998; R. Delort and F. Walter, *Histoire de l'environnement européen,* Paris: Presses Universitaires de France, 2001.

On the history of vacationing, P. Battilani, *Vacanze di pochi vacanze di tutti. L'evoluzione del turismo europeo,* Bologna: il Mulino, 2001.

3.

On the growth of Italian industry, V. Zamagni, *Dalla periferia al centro. La seconda rinascita economica dell'Italia (1861–1981),* Bologna: il Mulino, 2003. On the spread of electric power, R. Giannetti, *La conquista della forza. Risorse, tecnologia ed economia nell'industria elettrica italiana (1883-1940),* Milano: Angeli, 1985.

On pollution and the effects of sanitary legislation, Neri Serneri, *Incorporare la natura. Storie ambientali del Novecento,* cit.

Some case studies can be found in S. Adorno and S. Neri Serneri (eds), *Industrie, ambiente e territorio. Per una storia ambientale delle aree industriali in Italia,* Bologna: il Mulino, 2009, especially the Introduction, pp. 13–31, and the essays by M. Ruzzenenti, 'Industrie urbane. La 'Caffaro' di Brescia', pp. 113–131, G. Zucconi, 'Marghera e la scommessa industriale di Venezia', pp. 133–148, A. Ciuffetti, 'Industrializzazione e territorio nella conca ternana, 1884–2004', pp. 149–166, and G. Corona, 'Industrialismo e ambiente: le molte identità di Bagnoli', pp. 189–211. An important book on this theme is P.P. Poggio and M. Ruzzenenti (eds), *Il caso italiano. Industria, chimica e ambiente,* Milano: Jaca book,

2012; see especially the essays by S. Barca, 'Il capitalismo nelle vallate. Acque e industrie nell'Italia dell'Ottocento', pp. 39–74, N. Nicolini, 'Le lavorazioni chimiche nell'Ottocento in Italia e l'ambiente', pp. 75–97, P.P. Poggio, 'L'Acna e la Valle Bormida', pp. 123–171, S. Grassi, 'La Rumianca di Pieve Vergonte (1915–2012)', pp. 173–202, F. Nunnari, 'Il 'nucleo di industrializzazione Valle Del Sacco. Un rischioso tentativo di sviluppo', pp. 203–223, and M. Rizzenenti, 'La storia controversa del piombo tetraetile', pp. 225–251. On the conflict between the industry and the peasant world, P.P. Poggio, 'Resistenza contadina', *Lo Straniero* **141** (2012). On the relationship between water and industrialisation, S. Barca, *Enclosing Water. Nature and Political Economy in a Mediterranean Valley, 1796–1915*, Cambridge: The White Horse Press, 2010.

On land issues in the southern Italian debate, P. Bevilacqua, *Breve storia dell'Italia meridionale dall'Unità a oggi*, Roma: Donzelli, 1993.

On water resources and hydroelectric and irrigation projects, G. Barone, *Mezzogiorno e modernizzazione: elettricità, irrigazione e bonifiche nell'Italia contemporanea*, Torino: Einaudi, 1986. On Francesco Saverio Nitti, id., 'Francesco Saverio Nitti', in *Dizionario Biografico degli italiani*, www.treccani.it/enciclopedia/francesco-saverionitti_(Dizionario-biografico).

4.

On the history of ecological thought, D. Worster, *Nature's Economy. A History of Ecological Ideas*, Cambridge: Cambridge University Press, 1985 (First published 1977 by Sierra Club Books).

On the history of early nature defence movements in Italy, L. Piccioni, *Il volto amato della Patria. Il primo movimento per la protezione della natura in Italia 1880-1934*, Camerino: Università degli studi di Camerino, 1999; id., 'La natura come posta in gioco. La dialettica tutela ambientale-sviluppo turistico nella "regione dei parchi"', in M. Costantini and C. Felice (eds), *Storia d'Italia. Le Regioni dall'Unità a oggi. L'Abruzzo*, Torino: Einaudi, 2000; id., 'Paesaggio della Belle Epoque. Il catalogo delle bellezze naturali d'Italia 1913–1926', in P.P. Poggio and M. Ruzzenenti (eds), *Il caso italiano. Industria, chimica e ambiente*, Milano: Jaca book, 2012, pp. 99–121; J. Sievert, *The Origins of Nature Conservation in Italy*, Bern: Peter Lang, 2000; G. Della Valentina, *Storia dell'ambientalismo in Italia. Lo sviluppo sostenibile*, Milano: Bruno Mon-

Bibliography

dadori, 2011. See also A.F. Saba, 'Cultura, natura, riciclaggio. Il fascismo e l'ambiente dal movimento ruralista alle necessità autarchiche', in A.F. Saba and E. Mayer (eds), *Storia ambientale. Una frontiera storiografica*, Milano: Teti, 2001, pp. 63–100; M. Armiero, *Le montagne della patria. Natura e nazione nella storia d'Italia. Secoli XIX e XX*, Torino: Einaudi, 2013 (first published as *A Rugged Nation: Mountains and the Making of Modern Italy*, Cambridge: The White Horse Press, 2011) .

Chapter Three

1.

On world environment in the twentieth century, J.R. McNeill, *Something New under the Sun. An Environmental History of the Twentieth Century World*, New York: Norton and Company, 2000; F. Paolini, *Breve storia dell'ambiente nel Novecento*, Roma: Carocci, 2009.

On the energy transition, P. Malanima, *Le energie degli italiani. Due secoli di storia*, Milano: Bruno Mondadori, 2013.

On demographic changes, Svimez, *150 anni di statistiche italiane: Nord e Sud, 1861-2011*, Bologna: il Mulino, 2011.

On the transformation of cities and its effects on the ecosystemic balance, I used the mapping proposed by L. Gambi, 'I valori storici dei quadri ambientali', in *Storia d'Italia, I: I caratteri originali*, Torino: Einaudi, 1972.

On the environmental implications of industrialisation in the second postwar period, S. Neri Serneri, 'L'impatto ambientale dell'industria, 1950–2000. Risorse e politiche', in S. Adorno and S. Neri Serneri (eds), *Industria, ambiente e territorio. Per una storia ambientale delle aree industriali in Italia*, Bologna: il Mulino, 2009, pp. 33–86. In the same volume: R. Tolaini, 'Il peso dell'acciaio. Siderurgia e ambiente a Genova, 1950–2005', pp. 87–112; F. Paolini, 'Industria diffusa e inquinamento nell'area fiorentino-pratese, 1946–2001', pp. 167–187; M.G. Rienzo, 'Manfredonia tra sviluppo industriale e oltraggio ambientale', pp. 213–236; S. Ruju, 'Il petrolchimico di Porto Torres negli anni della Stir, 1957–1977', pp. 237–266; S. Adorno, 'L'area industriale siracusana e la crisi ambientale degli anni Settanta', pp. 267–316; M. Nucifora, 'Pianificazione e politiche per l'ambiente. Le aree industriali italiane nel secondo Novecento', pp. 317–338.

Gabriella Corona

On the Italian and European juridical debate on the environment, R. Giuffrida (ed.), *Diritto europeo dell'ambiente,* Torino: Giappichelli, 2012, especially the essay by A. Rizzo, 'L'affermazione di una politica ambientale dell'Unione Europea. Dall'Atto Unico Europeo al Trattato di Lisbona', pp. 3–16; L. Scichilone, *L'Europa e la sfida ecologica. Storia della politica ambientale europea (1969–1998),* Bologna: il Mulino, 2008. See also E. Rook Basile, S. Carmignani and N. Lucifero, *Strutture agrarie e metamorfosi del paesaggio. Dalla natura delle cose alla natura dei fatti,* Milano: Giuffrè, 2010, especially S. Carmignani's essay 'Paesaggio, agricoltura e territorio. Profili pubblicistici', pp. 1–97.

On the response of public institutions to environmental problems, S. Pinna, *La protezione dell'ambiente. Il contributo della filosofia, dell'economia e della geografia,* Milano: Angeli, 1995. For a complete overview of the legislation on pollution from the aftermath of World War II onward, G. Amendola, *Inquinamenti. Gli elementi essenziali di normativa e giurisprudenza in materia di acqua, rumore, rifiuti,* Roma: Epe libri, 2000.

For case studies on the repercussions of urban growth on the environment in the second postwar period, G. Corona and S. Neri Serneri (eds), *Storia e ambiente. Città, risorse e territori nell'Italia contemporanea,* Roma: Carocci, 2007, especially the essays by F. Paolini, 'I territori dello sviluppo. L'area fiorentino-pratese (1946–95)', pp. 179–194, S. Adorno, 'Il polo industriale di Augusta-Priolo. Risorse e crisi ambientale (1949–2000)', pp. 195–217, and S. Bartoletto, 'L'energia delle città. Percorsi di ricerca muovendo dal caso di Napoli', pp. 218–233.

On the transformation of Italian agriculture after World War II, G. Corona and G. Massullo, 'La terra e le tecniche. Innovazioni produttive e lavoro agricolo nei secoli XIX e XX', in P. Bevilacqua (ed.), *Storia dell'agricoltura italiana in età contemporanea, I: Spazi e paesaggi,* Venezia: Marsilio, 1989, pp. 353–449. On the transformations of the agrarian landscape over the last sixty years and the increasing gap between deteriorated mountain areas and plains given over to specialised crops, G. Barbera, *Tuttifrutti. Viaggio tra gli alberi da frutto mediterranei, fra scienza e letteratura,* Milano: Mondadori, 2007.

On the industrialisation of agriculture and animal husbandry and its effects on the environment, P. Bevilacqua, *La mucca è savia. Ragioni storiche della crisi alimentare europea,* Roma: Donzelli, 2002; E. Bernardi,

Bibliography

Il mais "miracoloso". Storia di un'innovazione tra politica, economia e reli-gione, Roma: Carocci, 2014. On industrial agriculture and the consumption of water resources: E. D'Angelis and A. Irace, *Il valore dell'acqua. Chi la gestisce, quanta ne consumiamo e come possiamo salvarla*, Milano: Dalai, 2011.

For data from epidemiological studies, see Sentieri, 'Studio epidemiologico nazionale dei territori e degli insediamenti esposti a rischio di inquinamento. Risultati', *Epidemiologia e Prevenzione* **35**/5–6 (2011), Suppl. 4: 1-204; or online at www. epiprev.it/pubblicazione/epidemiol-prev-2011-35-5-6-suppl-4.

2.

On land reclamation in general, P. Bevilacqua and M. Rossi-Doria, *Le bonifiche in Italia dal '700 ad oggi*, Roma, Bari: Laterza, 1984. On land reclamation in the context of Fascist politics, S. Lupo. *Il fascismo. La politica in un regime totalitario*, Roma: Donzelli, 2005.

Fascist period literature on land reclamation includes A. Serpieri, *La politica agraria in Italia e i recenti provvedimenti legislativi*, Piacenza: Federazione dei Consorzi Agrari, 1925; id., *La bonifica integrale*, Roma: Istituto Poligrafico dello Stato, 1930.

On extraordinary policy actions, L. D'Antone, 'L'"interesse straordinario" per il Mezzogiorno (1943–60)', *Meridiana* **24** (1995) 17–64; id. (ed.), *Radici storiche ed esperienza dell'intervento straordinario nel Mezzogiorno*, Roma: Bibliopolis, 1996; id., 'Il governo dei tecnici. Specialismi e politica nell'Italia del Novecento', *Meridiana* **38–39** (2000): 101–125; G. Barone, 'Stato e Mezzogiorno (1943–60). Il "primo tempo" dell'intervento straordinario', in *Storia dell'Italia repubblicana, I: La costruzione della democrazia. Dalla caduta del fascismo agli anni cinquanta*, Torino: Einaudi, 1994, pp. 293–409. On the political management of extraordinary policy actions in the Italian South, see S. Zoppi, *Il Mezzogiorno di De Gasperi e Sturzo (1944–1959)*, Soveria Mannelli: Rubbettino, 2003.

On the industrialisation of the South promoted by the Cassa per il Mezzogiorno, see P. Bevilacqua, *Breve storia dell'Italia meridionale dall'Ottocento ad oggi*, Roma: Donzelli, 1993, especially pp. 102–106. See also G. Sapelli, *Storia economica dell'Italia contemporanea*, Milano: Bruno Mondadori, 1997.

Gabriella Corona

On the reconquest of the sea, P. Frascani, *Il mare*, Bologna: il Mulino, 2008. On the multiplication of *marine* in Calabria, see P. Bevilacqua, 'Uomini, terre, economie', in *Storia d'Italia. Le Regioni dall'Unità a oggi, La Calabria*, Torino: Einaudi, 1985, pp. 142–163; G. Corona, 'La grande svolta negli assetti del territorio: bonifiche, rinascita delle pianure, urbanesimo', in *Storia della Calabria, Il Novecento*, Roma, Bari: Laterza, 2001.

On the spread of malaria and its effects, P. Corti, 'Malaria e società contadina nel Mezzogiorno', in F. Della Peruta (ed.), *Storia d'Italia, Annali 7: Malattia e medicina*, Torino: Einaudi, 1984, pp. 635–678; P. Tino, 'Malaria e modernizzazione in Italia dopo l'Unità', *I frutti di Demetra* **8** (2005): 27–37.

On agrarian reform, G. Massullo, 'Contadini. La piccola proprietà coltivatrice nell'Italia contemporanea', in P. Bevilacqua (ed.), *Storia dell'agricoltura italiana in età contemporanea, II: Uomini e classi*, Venezia: Marsilio, 1990, pp. 5–43.

For a critique of the deviant aspects of extraordinary policy actions in the Italian South, A. Becchi, 'Opere pubbliche', *Meridiana* **9** (1990) 223–243; C. Trigilia, *Sviluppo senza autonomia. Effetti perversi delle politiche nel Mezzogiorno*, Bologna: il Mulino, 1992.

3.

On the transformations of the Italian city in the second postwar period, C. De Seta, *Città, territorio e Mezzogiorno in Italia*, Torino: Einaudi 1977; V. De Lucia, *Se questa è una città. La condizione urbana nell'Italia contemporanea*, Roma: Donzelli, 2005; P.L. Cervellati, *La città bella*, Bologna: il Mulino, 1991; G. Dematteis, 'Gli anni del "miracolo economico": il territorio polarizzato', in *Storia dell'Italia repubblicana, II.1: La trasformazione dell'Italia: sviluppo e squilibri. Politica, economia, società*, Torino: Einaudi, 1995.

On urban-planning studies in general, E. Salzano, *Fondamenti di urbanistica: la storia e la norma*, Roma, Bari: Laterza, 2003. On town-planning in Italy after World War II, see G. Campos Venuti, *Amministrare l'urbanistica*, Torino: Einaudi, 1967; id., *Urbanistica e austerità*, Milano: Feltrinelli, 1978; L. Benevolo, *La città e l'architetto*, Roma, Bari: Laterza, 1984; G. Campos Venuti and F. Oliva (eds), *Cinquant'anni di urbanistica in Italia, 1942–1992*, Roma, Bari: Laterza, 1993.

Bibliography

On the Sullo Act, A. Becchi, 'La legge Sullo sui suoli', *Meridiana* **29** (1997) 107–134; V. De Lucia, *Nella città dolente. Mezzo secolo di scempi, condoni e signori del cemento dalla sconfitta di Fiorentino Sullo a Silvio Berlusconi*, Roma: Castelvecchi rx, 2013. On the relationship between urban planning and social conflict, A. Perelli (ed.), *La città fabbrica: contributi per una storia di classe del territorio*, Milano: Clup, 1970; A. Belli (ed.), *Città e territorio: pianificazione e conflitto*, Napoli: Coop. ed. Economia e Commercio, 1974; F. Indovina (ed.), *Dal blocco dei fitti all'equo canone. Il conflitto tra proprietari ed inquilini e le mediazioni delle forze politiche*, Venezia: Marsilio, 1977; A. Daolio (ed.), *Le lotte per la casa in Italia. Milano, Torino, Roma, Napoli*, Milano: Feltrinelli, 1974; M. Revelli, 'Movimenti sociali e spazio politico', in *Storia dell'Italia repubblicana, II.2: La trasformazione dell'Italia: sviluppo e squilibri. Istituzioni, movimenti, culture*, Torino: Einaudi, 1995, pp. 385–476. On the history of urban regeneration and the protection of historical centres, S. Muratori, *Studi per una operante storia urbana di Venezia*, Roma: Istituto Poligrafico dello Stato, 1960; P. Ceccarelli and F. Indovina, *Risanamento e speculazione nei centri storici*, Milano: Angeli, 1977; F. Ciccone (ed.), *Recupero e riqualificazione urbana*, Milano: Giuffrè, 1984. On urban regeneration and the protection of historical centres seen as a form of metabolism of places, see also R. D'Arienzo and C. Younès (eds), *Pour une écologie des milieux habités. Recycler l'urbain*, Genève: MetisPresses, 2014.

4.

On Italian environmentalism, its characteristics and its history in this period, E.H. Meyer, *I pionieri dell'ambiente. L'avventura del movimento ecologista italiano. Cento anni di storia*, Milano: Carabà, 1995; A. Poggio, *Ambientalismo*, Milano: Bibliografica, 1996; R. Della Seta, *La difesa dell'ambiente in Italia. Storia e cultura del movimento ecologista*, Milano: Angeli, 2002; S. Luzzi, *Il virus del benessere. Ambiente, salute, sviluppo nell'Italia repubblicana*, Roma, Bari: Laterza, 2009; G. Della Valentina, *Storia dell'ambientalismo italiano. Lo sviluppo insostenibile*, Milano: Bruno Mondadori, 2011; S. Neri Serneri, *Culture e politiche dei movimenti ambientalisti*, in F. Lussana and G. Marramao (eds), *L'Italia repubblicana nella crisi degli anni Settanta, II: Culture, nuovi soggetti, identità*, Soveria Mannelli: Rubbettino, 2003, pp. 367–400.

On the relationship between the student movement of the 1960s

Gabriella Corona

and the rise of political ecology in Italy, C. Papa, 'Alle origini dell'ecologia politica in Italia. Il diritto alla salute e all'ambiente nel movimento studentesco', in Lussana and Marramao (eds), *L'Italia repubblicana nella crisi degli anni Settanta*, II, cit., pp. 401–431.

On Antonio Cederna, F. Erbani, *Antonio Cederna. Una vita per la città, il paesaggio, la bellezza*, Morciano di Romagna: Biblioteca del Cigno, 2012.

On the report of the Club di Roma, D. Meadows et al., *The Limits to Growth*, New York: Universe books, 1972. On its repercussions for the Italian public debate, L. Piccioni and G. Nebbia, 'I limiti dello sviluppo in Italia. Cronache di un dibattito 1971–74', in *Fondazione Luigi Micheletti. I quaderni di Altronovecento* **1** (2011): 5–53.

On different uses of the term 'development', G. Rist, *The History of Development from Western Origins to Global Faith*, London, New York: Zed Books Ltd, 1997.

On the relationship between environmentalism and the Left, S. Gentili, *Ecologia e sinistra. Un incontro difficile*, Roma: Editori riuniti, 2002; P. Pelizzari, 'Sviluppo e ambiente nel dibattito della sinistra', *Italia Contemporanea* **247** (2007): 253–269. On health in factories, F. Carnevale and A. Baldasseroni, *Mal da lavoro. Storia della salute dei lavoratori*, Roma, Bari: Laterza, 1999.

On Enrico Berlinguer and austerity politics, E. Berlinguer, *Austerità. Occasione per trasformare l'Italia*, Roma: Editori riuniti, 1977; P. Della Seta and E. Salzano, *L'Italia a sacco*, Roma: Editori riuniti, 1993; G. Fiori, *Vita di Enrico Berlinguer*, Roma, Bari: Laterza, 2004; F. Barbagallo, *Enrico Berlinguer*, Roma: Carocci, 2006.

On Laura Conti, P. Pelizzari, 'Inquinamenti, industrialismo e impegno ambientale nell'attività di Laura Conti', in P.P. Poggio and M. Ruzzenenti (eds), *Il caso italiano. Industria, chimica e ambiente*, Milano: Jaca book, 2012, pp. 449–483.

On Giorgio Nebbia, P.P. Poggio, 'Intervista a Giorgio Nebbia', in Poggio and Ruzzenenti (eds), *Il caso italiano. Industria, chimica e ambiente*, cit., pp. 359–372.

On Seveso, A. Cutrera, G. Pastorelli and B. Pozzo (eds), *Seveso trent'anni dopo: la gestione del rischio industriale*, Milano: Giuffrè, 2006; B. Ziglioli, *La mina vagante. Il disastro di Seveso e la solidarietà nazionale*, Milano: Angeli, 2010.

Bibliography

Chapter Four

1.

On the legal and institutional context, G. Gisotti and M. Benedini, *Il dissesto idrogeologico. Previsione, prevenzione e mitigazione del rischio*, Roma: Carocci, 2000. For an historiographical perspective on hydrogeological risk, W. Palmieri, 'Dissesto e disastri idrogeologici nell'Italia unita', in G. Corona and P. Malanima (eds), *Economia e ambiente in Italia dall'Unità a oggi*, Milano: Bruno Mondadori, 2012.

On negotiated urban planning, V. De Lucia, *Nella città dolente. Mezzo secolo di scempi, condoni e signori del cemento dalla sconfitta di Fiorentino Sullo a Silvio Berlusconi*, Roma: Castelvecchi rx, 2013.

On urban sprawl, M.C. Gibelli and E. Salzano (eds), *No Sprawl*, Firenze: Alinea, 2006.

On green belts, see the classic book by F.J. Osborn, *Green-Belt Cities* (1946), London: Adams and Mackay, 1969. See also E. Howard, *The Garden City Movement*, London: Garden City Association, 1906, as well as P. Sica, *Storia dell'Urbanistica, III: Novecento*, Roma, Bari: Laterza, 1985, pp. 7–45, and C. Doglio, *La città giardino (1953)*, Roma: Gangemi, 1985.

On land consumption, A. di Gennaro, F.P. Innamorato, N. Filippi, F. Malucelli and B. Guandaloni, 'Come è cambiato il nostro territorio. Dinamiche di uso del suolo nei paesaggi italiani tra il 1990 e il 2006', *Territori* 3 (2011): 59–69; A. di Gennaro and F.P. Innamorato, *La grande trasformazione. Il territorio rurale della Campania 1960-2000*, Napoli: Clean, 2005. On the Italian coasts, F. Pratesi, *Storia della natura in Italia*, Soveria Mannelli: Rubbettino, 2011. See also S. Settis, *Paesaggio Costituzione cemento. La battaglia per l'ambiente contro il degrado civile*, Torino: Einaudi, 2010.

On metropolitan cities, volume 80 of *Meridiana* (2014), especially M.C. Gibelli, 'Milano città metropolitana fra deregolazione e nuova progettualità': 41–64; C. Carminucci, S. Casucci and G. Frisch, 'Roma, una città metropolitana in crescita e trasformazione': 77–104; A. di Gennaro, 'Per una storia dell'ecosistema metropolitano di Napoli': 105–124; V. De Lucia, 'La città metropolitana di Napoli. Poteva essere un'occasione di riscatto': 125–142; G. Macri, 'Profili istituzionali del dibattito sulla città metropolitana': 173–196.

Gabriella Corona

On motor vehicle traffic as a general environmental problem, J.R. McNeill, *Something New under the Sun. An Environmental History of the Twentieth Century World*, New York: Norton and Company, 2000. On the case of Italy, F. Paolini, *Un paese a quattro ruote. Automobili e società in Italia*, Venezia: Marsilio, 2005; S. Bartoletto, 'Le trasformazioni dei flussi energetici delle città italiane in età contemporanea', paper presented at the conference *Ambiente e storia. Risorse, città e territori nell'Italia contemporanea*, Siena, 9–10 December 2005; id., 'Energia e ambiente in Europa (1800–2010)', *Rivista di studi sulla sostenibilità* 1 (2011): 59–78; S. Bartoletto and M. del Mar Rubio Varas, 'Energy transition and CO2 emissions in Southern Europe: Italy and Spain (1861-2000)', *Global Environment* 2 (old series, 2008): 47–81; Legambiente, *Dossier inquinamento atmosferico*, Roma, 20 November 2001 and id., 'Dossier ecosistema urbano', 2003, at www.legambiente.com.

2.

On the rise of political environmentalism, F. Giovannini (ed.), *Culture della sinistra e culture verdi. La sfida della rivoluzione ambientale*, Roma: Datanews, 1994; D. Della Porta and M. Diani, *Movimenti senza protesta? L'ambientalismo in Italia*, Bologna: il Mulino, 2004; S. Luzzi, *Il virus del benessere. Ambiente, salute, sviluppo nell'Italia repubblicana*, Roma, Bari: Laterza, 2009; G. Della Valentina, *Storia dell'ambientalismo in Italia. Lo sviluppo insostenibile*, Milano: Bruno Mondadori, 2011.

On the bases of ecological economy, E. Tiezzi and N. Marchettini, *Che cos'è lo sviluppo sostenibile? Le basi scientifiche della sostenibilità e i guasti del pensiero unico*, Roma: Donzelli, 1999. On the history of sustainability, G. Senatore, *Storia della sostenibilità. Dai limiti della crescita alla genesi dello sviluppo*, Milano: Angeli, 2013. On contradictions in the concept of sustainable development, *Défaire le développement, refaire le monde*, Parangon Paris : L'Aventurine, 2003.

On green parties in Europe and the world, F. Paolini, 'I partiti politici ecologisti dal "successo" al riflusso (1972–2008). Appunti per una storia dell'ambientalismo politico', *I frutti di Demetra* 18 (2008): 35–48. On actions carried out by the world of associations, A. Fiorillo, M. Fratoddi and S. Venneri (eds), *Ricomincio da trenta. Sfide, battaglie e buone idee per il futuro dell'Italia*, Morciano di Romagna: Biblioteca del Cigno, 2010.

Bibliography

The notion of 'great political depression' is borrowed from A. Mastropaolo, *La mucca pazza della democrazia. Nuove destre, populismo, antipolitica,* Torino: Bollati Boringhieri, 2005.

3.

On waste in an historical perspective, E. Sori, *Il rovescio della produzione. I rifiuti in età pre-industriale e paleo tecnica,* Bologna: il Mulino, 1999; id., *La città e i rifiuti. Ecologia urbana dal Medioevo al primo Novecento,* Bologna: il Mulino, 2001; R. Sansa, 'I rifiuti e la storia ambientale. Un'introduzione', *Storia urbana* **112** (2006): 7–16; P. Bevilacqua, 'La metamorfosi dissipativa della natura', *Meridiana* **64** (2009): 27–39; C. Montalbetti and E. Sori (eds), *Quel che resta di un bene. Breve storia della raccolta differenziata e del riciclaggio di carta e cartone,* Bologna: il Mulino, 2011; A. Massarutto, *I rifiuti. Come e perché sono diventati un problema,* Bologna: il Mulino, 2009.

On waste emergencies as a national problem, D. Fortini, 'Ormai sono vent'anni che il paese è in emergenza rifiuti', conversation with G. Corona, *Meridiana* 64 (2009): 41–69; G. Corona and D. Fortini, *The Problem of Waste Disposal in a Large European City. Garbage in Naples,* New York: Mellen Press, 2012; G. Corona, 'Nuove e vecchie emergenze', *Ecoscienza* **1** (2011): 18–19.

For an ecological approach to waste disposal issues, see G. Viale, *Azzerare i rifiuti. Vecchie e nuove soluzioni per una produzione e un consumo sostenibili,* Torino: Bollati Boringhieri, 2008; A. Cavaliere, *Il mucchio selvaggio. Per un'ecologia dei rifiuti,* Napoli: l'Ancora del Mediterraneo, 2006; L. Venturi (ed.), *Pianeta rifiuti,* Montepulciano: Le Balze, 2006, Series: I quaderni di Legambiente.

The cited data are those of the Istituto Superiore per la Protezione e la Ricerca ambientale (ISPRA), *Rapporto Rifiuti urbani,* 2013, online at www.isprambiente.gov.it/it/pubblicazioni/ rapporti/rapporto-rifiuti-urbani-edizione-2013; id., *Rapporto Rifiuti Urbani,* 2014, online at www.isprambiente.gov.it/ files/pubblicazioni/rapporti/rapportorifiutiUrbani2014-webpdf.

4.

On Ecomafias, much information is to be found in the acts of committees on the waste cycle and illicit activities connected to it in the 12th, 13th,

136

Gabriella Corona

14th, 15th and 16th legislature. Another source is *Commissione parlamentare antimafia, Camorra e politica. Relazione approvata dalla Commissione il 21 dicembre 1993*, Roma, Bari: Laterza, 1994. Important annual reports have been produced by the Osservatorio Ambiente e Legalità di Legambiente, notably: *Rapporto Ecomafia 2008. I numeri e le storie della criminalità ambientale*, Milano: Edizioni ambiente, 2008; *Ecomafia 2010. Le storie e i numeri della criminalità ambientale*, Milano: Edizioni ambiente, 2010; *Ecomafia 2013. Le storie e i numeri della criminalità ambientale*, Milano, Edizioni ambiente: 2013; see also Legambiente, *Dossier inquinamento atmosferico 2001* and *Dossier ecosistema urbano 2003*, at www.legambiente.com

On Ecomafias, again, see V. Mete, *Fuori dal Comune. Lo scioglimento delle amministrazioni locali per infiltrazioni mafiose*, Acireale-Roma: Bonanno, 2009; P. Berdini, *Breve storia dell'abuso edilizio in Italia dal ventennio fascista al prossimo futuro*, Roma: Donzelli, 2010; R. Sciarrone (ed.), *Alleanze nell'ombra. Mafie ed economie locali in Sicilia e nel Mezzogiorno*, Roma: Donzelli, 2011; id. (ed.), *Mafie del Nord. Strategie criminali e contesti locali*, Roma: Donzelli, 2014.

On the case of Campania, I. Sales (ed.), *Rapporto sulla camorra 1991, Comitato regionale Pds Campania*; F. Barbagallo, *Napoli fine Novecento. Politici, camorristi, imprenditori*, Torino: Einaudi, 1997; R. Saviano, *Gomorra. Viaggio nell'impero economico e nel sogno di dominio della camorra*, Milano: Mondadori, 2006; M. Andretta (ed.), 'Traffico illegale di rifiuti tossici: un caso solo campano?', interview with Donato Ceglie, *I frutti di Demetra* 16 (2008): 49–59; R. Capacchione, *L'oro della camorra. Come i boss casalesi sono diventati ricchi e potenti manager*, Milano: Rizzoli, 2008; M. Andretta, 'Da Campania felix a discarica. Le trasformazioni in Terra di Lavoro dal dopoguerra a oggi', *Meridiana* 64 (2009): 87–113; G. Gribaudi (ed.), *Traffici criminali. Camorra, mafie e reti internazionali dell'illegalità*, Torino: Bollati Boringhieri, 2009; M. Braucci and S. Laffi (eds), *Terre in disordine. Racconti e immagini della Campania di oggi*, Roma: Minimum Fax, 2009; the 'Ecocamorre' volume of *Meridiana* 73–74 (2012), especially G. Corona and R. Sciarrone, 'Il paesaggio delle ecocamorre': 13–34; I. Sales, 'La questione rifiuti e la camorra': 63–79; R. Cantone, 'I crimini contro il territorio': 81–88; D. Fortini, 'Rifiuti urbani e rifiuti speciali: i fattori strutturali delle ecocamorre': 89–102; V. Martone, 'La camorra nella governance del territorio': 13–34; G. Belloni, 'Camorra e criminalità ambientale in

Bibliography

Veneto': 103–131; A. Di Lorenzo, 'L'anticittà della camorra': 173–190; A. di Gennaro, 'Un piano per uscire da Gomorra': 191–202; L. Musella, 'I "confini" della camorra': 209–225.

On the Land of Fires, A. di Gennaro, 'Per una storia dell'ecosistema metropolitano di Napoli' and M. Demarco, 'La Terra dei fuochi? Un problema di rappresentanza?'*Meridiana* 80 (2014): respectively 105–124 and 221–227. See also A. di Gennaro, *La terra ferita. Cronistoria della Terra dei Fuochi*, Napoli: Clean, 2015. For a historiographical perspective, see P. Tino, *Campania felice? Territorio e agricolture prima della 'grande trasformazione'*, Catanzaro: Meridiana Libri, 1997.

On the environmental problems of Campania, I will limit myself to citing the essays in volume **42** of 2001 of *Meridiana*, 'Napoli sostenibile', as well as F. Ceci and D. Lepore, *Arcipelago Vesuviano. Percorsi e ragionamenti intorno a Napoli*, Lecce: Argo, 1997.

Conclusions

On the current situation of the environment in Italy, Legambiente, D. Bianchi and R. Della Seta (eds), *Ambiente in Europa. Economia verde: Italia-Germania è sempre 4 a 3?* (Ambiente Italia 2014. Rapporto annuale di Legambiente), Milano: Edizioni Ambiente, 2014.

On the latest transformations of the energy system, S. Bartoletto, 'I combustibili fossili in Italia dal 1870 ad oggi', *Storia economica* **2** (2005): 281–327; A. Cardinale and A. Verdelli, *Energia per l'industria in Italia: la variabile energetica dal miracolo economico alla globalizzazione*, Milano: Angeli, 2008; A. Clô, *Il rebus energetico*, Bologna: il Mulino, 2008; V. Pisano, *Lo sviluppo delle fonti rinnovabili in Italia*, lulu.com, 2008; P. Malanima, *Le energie degli italiani. Due secoli di storia*, Milano: Bruno Mondadori, 2013; M. Puttilli, *Geografia delle fonti rinnovabili. Energia e territorio per un'eco-ristrutturazione della società*, Milano: Angeli, 2014. For statistics, www.terna.it/default/Home/SiSTEMa_ElETTriCo/ Statistiche/dati_ statistici.aspx and www.gse.it/it/Statistiche/Pages/default.aspx, especially *GSE (Gestione Servizi Energetici), Rapporto statistico 2012. Impianti e fonti rinnovabili. Settore elettrico.*

The data on biological agriculture are from *Bioreport 2013. L'agricoltura biologica in Italia*, Roma, Rete Rurale Nazionale 2007–2013, 2013.

Gabriella Corona

On wine-and-food tourism, R. Pazzagli, *Il Buonpaese. Territorio e gusto nell'Italia in declino*, Pisa: Felici, 2013; the data are at www.coldiretti.it/News/Pagine/572----23-agosto-2013.aspx

On the green economy and overcoming the crisis, E. Zanchini, 'Uscire dalla crisi puntando sulla green economy', *Italianieuropei* **1** (2015), www.italianieuropei.it/it/italianieuropei-1-2015

INDEX

Index

Gabriella Corona

Index

Index

Index